The Socio-Economic Causes and Consequences of Desertification in Central Asia

NATO Science for Peace and Security Series

This Series presents the results of scientific meetings supported under the NATO Programme: Science for Peace and Security (SPS).

The NATO SPS Programme supports meetings in the following Key Priority areas: (1) Defence Against Terrorism; (2) Countering other Threats to Security and (3) NATO, Partner and Mediterranean Dialogue Country Priorities. The types of meeting supported are generally "Advanced Study Institutes" and "Advanced Research Workshops". The NATO SPS Series collects together the results of these meetings. The meetings are coorganized by scientists from NATO countries and scientists from NATO's "Partner" or "Mediterranean Dialogue" countries. The observations and recommendations made at the meetings, as well as the contents of the volumes in the Series, reflect those of participants and contributors only; they should not necessarily be regarded as reflecting NATO views or policy.

Advanced Study Institutes (ASI) are high-level tutorial courses intended to convey the latest developments in a subject to an advanced-level audience

Advanced Research Workshops (ARW) are expert meetings where an intense but informal exchange of views at the frontiers of a subject aims at identifying directions for future action

Following a transformation of the programme in 2006 the Series has been re-named and re-organised. Recent volumes on topics not related to security, which result from meetings supported under the programme earlier, may be found in the NATO Science Series.

The Series is published by IOS Press, Amsterdam, and Springer, Dordrecht, in conjunction with the NATO Public Diplomacy Division.

Sub-Series

A.	Chemistry and Biology	Springer
B.	Physics and Biophysics	Springer
C.	Environmental Security	Springer
D.	Information and Communication Security	IOS Press
E.	Human and Societal Dynamics	IOS Press

http://www.nato.int/science
http://www.springer.com
http://www.iospress.nl

Series C: Environmental Security

The Socio-Economic Causes and Consequences of Desertification in Central Asia

edited by

Roy Behnke
Macaulay Institute
Craigiebuckler, Aberdeen
UK

Published in cooperation with NATO Public Diplomacy Division

Proceedings of the NATO Advanced Research Workshop on
The Socio-Economic Causes and Consequences of
Desertification in Central Asia
Bishkek, Kyrgyzstan
June 2006

Library of Congress Control Number: 2008928514

ISBN 978-1-4020-8543-7 (PB)
ISBN 978-1-4020-8542-0 (HB)
ISBN 978-1-4020-8544-4 (e-book)

Published by Springer,
P.O. Box 17, 3300 AA Dordrecht, The Netherlands.

www.springer.com

Printed on acid-free paper

PREFACE

This book contains a selection of papers presented at the Advanced Research Workshop on 'The Socio-economic causes and consequences of desertification in Central Asia' held in Bishkek, Kyrgyzstan, in June 2006. The meeting provided a forum for twenty-six scientists from Central Asia and NATO countries to discuss the human dimensions of the desertification process. Papers presented to the meeting examined recent scientific evidence on the impact of desertification on livestock production, public health, and biodiversity, and contributed to the formulation of coherent national and regional policies for the management of watersheds, rangelands, and irrigated agriculture. The meeting was co-directed by Roy Behnke of the Macaulay Institute, UK, and by Lapas Alibekov of the Samarkand State University, Uzbekistan. Both the workshop and this subsequent publication have been financed by the NATO Scientific Affairs Division and we gratefully acknowledge this support. The Bishkek meeting was ably hosted by the Kyrgyz Sheep Breeders Association under the direction of Akylbek Rakaev who contributed substantially to the successful running of the workshop.

Deliberations at the workshop emphasized that policy failures at national level had promoted desertification within the region. These failures included:

- Environmentally destructive and wasteful agricultural technologies dating from the period of the Soviet Union
- Post-independence monopolisation of natural resources by national elites, which created poverty and forced poorer households to engage in unsustainable agricultural practices
- Ineptly formulated, unenforced and conflicting national land laws on natural resource ownership and control
- Scientifically uninformed national policy formulation, which failed to address existing and documented resource management problems.

Papers presented to the workshop examined these issues as challenges to regional environmental policy formulation, with respect to overgrazing by livestock, and in terms of a series of case studies of natural resource degradation and desertification control. The papers in this book are grouped under these three headings.

Part I Challenges to policy formulation

Chapter 1 by Ilka Lindt, *The Link between Desertification and Security in Central Asia*, assesses the importance of desertification for regional security in Central Asia. Lindt notes that all Central Asian countries have limited supplies of three basic natural resources - fertile land, clean water and a healthy environment. At least one of these resources, often two, is in short supply because they are very unevenly distributed within the region. The situation is exacerbated when high population growth produces densely populated areas (e.g. Fergana Valley) where poor rural populations depend for their livelihoods on the quality and availability of natural resources. Under these conditions, desertification poses a 'soft security challenge' that contributes to the economic marginalization of certain sectors of the rural population and intensifies crises caused by a complex array of factors. As a contributing element in political, social and economic instability, desertification is closely associated with a number of regional security concerns. Desertification affects food and health security when it contributes to malnutrition, increased incidence of disease, and child mortality. Desertification influences livelihood security because it may force people to leave their homes and migrate when degraded land can no longer support them. Desertification has an effect on social and national security when it encourages civil unrest. Finally desertification affects international security through cross-boarder migration and increased inter-ethnic tensions.

Chapter 2 by Lindsay Stringer, *From Global Environmental Discourse to Local Adaptations and Responses: A Desertification Research Agenda for Central Asia,* examines the suitability for Central Asia of current international approaches to addressing desertification. The UN Convention to Combat Desertification (UNCCD) was one of the key conventions emerging from the 1992 Conference on Environment and Development. The UNCCD highlighted three key features of its approach: decentralisation, participation, and the importance of local knowledge in anti-desertification activities. Stringer argues that there has been little reflection on the suitability of the UNCCD's decentralized, participatory approaches to natural resource management, for the specific socio-political circumstances of the transition states. It is unclear the extent to which local knowledge of land management was lost in Central Asia during the Soviet period, or whether the economic difficulties experienced over the past 16 years have led to new ways of thinking about people's relationships with their environments. It is also unclear whether decentralized approaches can be easily implemented in a society where public participation has been historically absent. Finally, as indicated by the dated reference lists at the end of the National Action Plans to combat desertification that have been prepared by Central Asian governments, officially recognized land degradation and desertification priorities are based on data from the Soviet period and lack contemporary empirical evidence. These questions form some of the key gaps in desertification studies and are in urgent need of improved understanding. This chapter outlines the early stages of a research project that aims to explore the links between the global environmental discourse on desertification and local socio-economic impacts and responses to land degradation in Central Asia.

Chapter 3, *Causes and Socio-Economic Consequences of Desertification in Central Asia,* by L. Alibekov and D. Alibekov argues that agriculture in Central Asia was mismanaged for decades under a centralized command economy. The continuing legacy of this period is severe land degradation in the form of soil salinization and erosion, elevated groundwater levels caused by poorly managed irrigation systems, the drying of the Aral Sea, and the chemical and nuclear pollution of water and soil. This chapter provides an overview of how these failures occurred and describes some of their negative social and economic impacts. The authors further argue that private, non-governmental activities alone are insufficient to address these complex problems. Desertification takes place within naturally delimited geographical and ecological systems; it must also be studied and prevented within the framework of these natural systems. The authors advocate a comprehensive approach to the economic development of an entire country or region based on systematic and coordinated scientific research. Scientific research therefore becomes an increasingly important component in the search for ways to sustainably develop vulnerable ecosystems. Given that a significant part of Central Asia's natural resources have been exhausted and the ecological situation continues to deteriorate, urgent action is required.

Part II Grazing systems and desertification

This section contains four chapters that analyze rangeland degradation in Turkmenistan and Kazakhstan, two countries that have pursued contrasting policies towards their pastoral sector in the post-Soviet period.

In Kazakhstan, the privatisation of land and livestock began in the early 1990s and was complete by the end of the decade. In Turkmenistan, on the other hand, rangelands and many livestock still remain state property, with pastoralists working within what amount to reformed and renamed Soviet farms. Changes in pastoral land use reflect these national policy differences. In Kazakhstan, the sudden and chaotic privatisation of livestock caused the loss of about three quarters of the national herd in the mid-1990s. With the collapse of rural livelihood systems, there were high levels of emigration from pastoral areas into towns or larger rural settlements, rural farmsteads and wells were abandoned and destroyed, and many remote seasonal pastures were unused. Around 1999 or 2000 these downward trends were halted and then reversed as flocks expanded for the first time in a decade and larger flock owners began to re-colonize isolated farmsteads and wells.

Turkmenistan represents the antithesis to Kazakhstan's radical reforms. Independent Turkmenistan operates a centralized agricultural economy modelled on farm reforms that were being implemented in the Soviet Union in the late 1980s, just prior to Turkmenistan becoming independent. Households are allowed to lease livestock from the state and private ownership of livestock is permitted, but there is no private ownership or leasing of pastures or water points. The slow pace of agricultural reform in Turkmenistan did not precipitate the catastrophic livestock losses that accompanied radical reform in Kazakhstan. The proportion of the national flock that is private or state owned is unclear, though official statistics state that well over half of all small ruminants are now in private hands and that livestock are now more numerous than at any other time in Turkmenistan's history.

Chapter 4, *Forage Distributions, Range Condition, and the Importance of Pastoral Movement in Central Asia – a Remote Sensing Study*, by Michael Coughenour, Roy Behnke, John Lomas and Kevin Price, uses satellite data to compare the environmental impacts of Kazakh and Turkmen pastoral policies. In Kazakhstan, there is evidence for the regeneration of some destocked seasonal pastures, while other seasonal pasture areas show little response to the withdrawal of grazing, perhaps because they were not as damaged as the Soviet authorities had asserted.

There is also evidence of increased degradation around large settlements, described in Chapter 5, *The Impact of Livestock Grazing on Soils and Vegetation around Settlements in Southeast Kazakhstan,* by I. Alimaev, C. Kerven, A. Torekhanov, R. Behnke, V. K. Smailov, V. Yurchenko, Zh. Sisatov and K. Shanbaev. The availability of grazing around settlements was not a problem in the mid-1990s when flocks were small, but has become a constraint as animal numbers rebound. Uncontrolled grazing and high stocking rates around settlements have produced both environmental degradation and diminished livestock performance despite overall declines in sheep numbers in the post-socialist period.

In Turkmenistan there is evidence of degradation in the form of reduced rain use efficiency around wells and permanent settlements. Chapter 4 discusses the botanical evidence to support these conclusions; Chapter 7, *Human and Natural Factors that Influence Livestock Distributions and Rangeland Desertification in Turkmenistan*, Roy Behnke, Grant Davidson, Abdul Jabbar and Michael Coughenour, examines the distributions of livestock populations that cause observed changes in vegetation. Larger water points attract more settlers, support more animals and produce significant changes in the composition and biomass of surrounding vegetation, though it is unclear whether these vegetation changes are caused primarily by grazing or by fire wood collecting.

Herd/flock dispersal and seasonal mobility limit the environment impact of livestock. This is one of the main results of a multi-country study of pastoralism in Inner Asia, the region lying to the east of Central Asia and consisting of Mongolia, western China and southern Siberia:

> The highest levels of degradation were reported in districts with the lowest livestock mobility; in general, mobility indices were a better guide to reported degradation levels than were densities of livestock. This pattern corresponded with the experience of local pastoralists. At six sites, locals explicitly associated pasture degradation with practices that limited the mobility of livestock.[1]

Research conducted in Kazakhstan supports these conclusions and is presented in Chapter 6, *Livestock Mobility and Degradation in Kazakhstan's Semi-Arid Rangelands*, by Carol Kerven, Kanat Shanbaev, Iliya Ilitch Alimaev, Aidos Smailov and Kazbek Smailov. This study also documents the reasons why the owners of small and large herds generally pursue different mobility and production strategies. Small-scale village-based livestock owners rely on their animals mainly for subsistence. Compared to large-scale owners, they gain higher rates of economic returns per head of animal owned. Large-scale owners have returned to moving their animals to distant pastures, and their animals are heavier as a consequence. These types of owners can achieve economies of scale, but they have high actual costs as a result of moving their animals.

Part III Case studies of resource degradation and desertification control

Chapter 8, *Land Reform in Tajikistan: Consequences for Tenure Security, Agricultural Productivity and Land Management Practices*, by Sarah Robinson, Ian Higginbotham, Tanya Guenther and Andrée Germain, examines the impact of land reform on agricultural productivity. Recent legislation in Tajikistan allows farmers to obtain access to heritable land shares for private use, but reform has been geographically uneven. In mountainous areas reform has led to privatization of arable land but farming households are often renters or shareholders in 'collectives' whose managers still have much control over the land. In productive areas of the lowlands there has been little distribution to households of land suitable for cultivation, and much farming is still conducted by labourers working on collective farms. Access to pasture is generally good but some remote pastures have been abandoned due to risks and costs association with travelling to them, increasing pressure on pastures near villages.

---

[1] D. Sneath 1998. State policy and pasture degradation in Inner Asia. *Science* 281, page 1148

These reforms have lead to the following patterns in agricultural productivity and investment in soil fertility:

- Agricultural productivity is highest on the most 'private' form of tenure regime which is that of the household plot. Yields are lowest on collective 'state enterprises'.
- The poorest farmers have the lowest rates of fertiliser application and the lowest crop yields and are the least likely to invest in sustainable soil management.
- Where cotton is grown, crop rotation is almost never used, and intensive farming combined with inefficient irrigation has led to high levels of soil salinity.
- The biggest land degradation threat comes from fuel burning which causes erosion and loss of pasture plants for grazing.
- High population growth and lack of alternatives to agriculture hinder farm consolidation and leave many households in a subsistence trap, unable to accumulate capital to make long term investments in their land.

In general, where distribution has occurred some households have prospered, yet many have been left landless or with insecure tenure. Poorer households generally have the least secure tenure arrangements, worst quality land and smallest land areas. Unsustainable use of soils is most likely to occur amongst these groups.

Chapter 9, *Israeli Experience in Prevention of Processes of Desertification,* by N. Orlovsky, describes a successful national program of desertification control. Since over 60% of Israel is occupied by the Negev Desert, measures to combat desertification were initiated at an early stage of the country's development and have intensified. In the agricultural sector, substantial savings have been achieved through technological improvements in irrigation methods, increasing the efficient use of water and effluents, promoting water recycling, minimizing pesticide use, advancing organic and greenhouse agricultures, and the development of new crops and innovative machinery. Scarcity of water, limited land resources, and lack of natural resources have led Israel to base its economy on technological advances that have considerably reduced the risk of desertification.

Chapter 10 by Marián Janiga reviews the *Potential Effects of Global Warming on Atmospheric Lead Contamination in the Mountains*. Atmospheric metal contamination, especially lead, responds to both the environmental factors and global warming parameters as a function of altitude. Seasonal fluctuations in the lead concentrations of foliar parts of alpine plants have been recorded with values higher in winter and early spring months than in summer months. The larger the snow-free catchments area and the warmer conditions will be in the mountains, the larger and earlier dispersal of lead to the surrounding sub-mountain regions may occur. Pb and Al concentrations in the alpine plants and vertebrates must be of concern in acidified habitats.

Chapter 11, *The Influence of Environmental Factors on Human Health in Uzbekistan*, by M. Shamsiyev and Sh. A. Khusinova documents the impact of agricultural mismanagement on human health. The incidence of communicative and tropical diseases is high in the Aral Sea region of Central Asia, effecting 60–300 per 10,000 people. High levels of chlorides in drinking water, the high incidence of heart disease, and increased levels of Ca and Mg leading to biliary and renal calculosis are all problems. Morbidity rates for biliary calculosis have increased ten fold, rates for chronic gastritis four fold, renal diseases eight fold and arthrosis and arthritis by five and seven times. The incidence of acute respiratory diseases in Karakalpakistan varies from 46% to 52%, with the incidence of tuberculosis having doubled in the last ten years so that it is now three times the rate elsewhere in Uzbekistan. Complications in pregnancy affect roughly two thirds of all women who are exposed to pesticides, with increases in pregnancy induced hypertension, gestational anaemia, spontaneous abortions and preterm deliveries.

Roy Behnke
Macaulay Institute
Aberdeen
UK

CONTENTS

CONTRIBUTORS

Davlat Alibekov
Department of Physical Geography and Geoecology, Samarkand State University, 703004, 15 University Blvd. Samarkand, Uzbekistan

Lapas Alibekov
Department of Physical Geography and Geoecology, Samarkand State University, 703004, 15 University Blvd. Samarkand, Uzbekistan

Ilya I. Alimaev
Kazakh Scientific Centre for Livestock and Veterinary Research, Dzandosov Str. 31, 480035 Almaty, Kazakhstan

Roy Behnke
Macaulay Institute, Aberdeen, AB15 8QH, Scotland, U.K.

Michael Coughenour
Natural Resource Ecology Laboratory, Colorado State University, Fort Collins, CO 80523-1499, U.S.A.

Grant Davidson
Macaulay Institute, Craigiebuckler, Aberdeen AB15 8QH, U.K.

Andrée Germain
Dept. of Epidemiology, University of Ottawa, 451 Smythe Rd, Ottawa, ON, K1H 85N, Canada

Tanya Guenther
International Medical Corps, 1313 L St. NW, Suite 220, Washington, DC 20005, USA

Ian Higginbotham
Stabilization and Reconstruction Task Force, Foreign Affairs and International Trade Canada, 125 Sussex Drive, Ottawa, ON, K1A 0G2, Canada

Abdul Jabbar
Mik.-10, Oguz Han Str., Proezd-4 D-11, Korp-2, Kv-28, Ashgabat, Turkmenistan

Marián Janiga
Institute of High Mountain Biology, Zilina University, SK 05956 Tatranská Javorina 7, Slovak Republic

Carol Kerven
Macaulay Institute, Craigiebuckler, Aberdeen AB15 8QH, UK

Sh. A. Khusinova
Samarkand State Medical Institute, A.Temur Str., 18, Samarkand, Uzbekistan

Ilka Lindt
Regional Coordination Office of the Convention Project to Combat Desertification, (GTZ), Orbita 1, House 40, 480043 Almaty, Kazakhstan

John Lomas
Kansas Applied Remote Sensing Program, University of Kansas Lawrence, Kansas, 66045 U.S.A.

N. Orlovsky
Department of Dryland Biotechnologies, The J. Blaustein Institutes for Desert Research, Ben-Gurion University of the Negev, Sede Boqer Campus 84990, Midreshet, Ben-Gurion, Israel

Kevin Price
Kansas Applied Remote Sensing Program, University of Kansas Lawrence, Kansas, 66045 U.S.A.

Sarah Robinson
La Cousteille, Saurat, 09400, France

A. M. Shamsiyev
Samarkand State Medical Institute, A.Temur Str., 18, Samarkand, Uzbekistan

Kanat Shanbaev
Kazakh Scientific Centre for Livestock and Veterinary Research, Dzandosov Str. 31, 480035 Almaty, Kazakhstan

Zheksinbai Sisatov
Kazakh Scientific Centre for Livestock and Veterinary Research,
Dzandosov Str. 31, 480035 Almaty, Kazakhstan

Aidos Smailov
Association of Oil and Gas Energy Sector in Kazakhstan, Almaty,
Kazakhstan

Kazbek Smailov
Kazakh Scientific Centre for Livestock and Veterinary Research,
Dzandosov Str. 31, 480035 Almaty, Kazakhstan

Lindsay C. Stringer
Sustainability Research Institute, School of Earth and Environment,
University of Leeds, LS2 9JT, UK

Aibyn Torekhanov
Kazakh Scientific Centre for Livestock and Veterinary Research,
Dzandosov Str. 31, 480035 Almaty, Kazakhstan

Vladimir Yurchenko
Kazakh Scientific Centre for Livestock and Veterinary Research,
Dzandosov Str. 31, 480035 Almaty, Kazakhstan

PART I

CHALLENGES TO POLICY FORMULATION

CHAPTER 1

THE LINK BETWEEN DESERTIFICATION AND SECURITY IN CENTRAL ASIA

DESERTIFICATION AND SECURITY

ILKA LINDT*

Regional Coordination Office of the Convention Project to Combat Desertification, (GTZ), Orbita 1, House 40, 480043 Almaty, Kazakhstan

Abstract: This paper examines the linkage between desertification and security, focusing on Central Asia. The first part of the paper discusses the theoretical background and the second half concentrates on the situation in Central Asia. The paper concludes with recommendations for policy options to prevent desertification induced threats to security in the region.

Keywords: Security concepts; national security; regional security; desertification; Central Asia; security policy; environmental security

1. Definitions and aspects of security

The term "security" is a very general civil understanding of a value, a universally applied normative concept, which has different meanings. Basically it is a political value which is strongly linked with individual and civil value systems (Brauch, 2005a, 2005b). Security can be generally defined as the absence of a threat towards a commonly understood value.

The traditional concepts of security based on national sovereignty and territorial security were reviewed during the 1990s. A **broader definition** includes non-traditional threats into the characterization of security (Kepner and Rubio, 2003; Kepner, 2003a). A change could be observed from solely military threats towards an inclusion of various other challenges: less urgent and non-violent, soft security problems such as migration and

* To whom correspondence should be addressed. Ilka Lindt, Regional Coordination Office of the Convention Project to Combat Desertification, (GTZ), Orbita 1, House 40, 480043 Almaty, Kazakhstan; e-mail: lindt.gtzccd@web.de

R. Behnke (ed.), The Socio-Economic Causes and Consequences of Desertification in Central Asia, 3–11.

TABLE 1. Broadened security concepts

Label	Reference object	Value at risk	Sources of threat
National security	The state	Territorial integrity	State, substate actors
Societal security	Societal groups	National identity	Nations, migrants
Human security	Individuals, mankind	Home, group survival	Nature, state, globalization
Environmental security	Ecosystem	Sustainability	Mankind
Gender security	Gender relations, indigenous people, minorities	Equality, identity, solidarity	Patriarchy, totalitarian institutions (governments, churches, elites), intolerants

Source: Brauch, H.G. 2003

starvation (Brauch, 2005a). The fundamental changes of the international political setting observed during the 1990s resulted in new vulnerabilities and risks which are differently understood and interpreted (Brauch, 2005d). Simultaneously, the concept of **security was deepened** in the sense of a shift from state (national security) towards the individual (human security) – differentiations are nowadays made between international, regional, national and civil security (Brauch, 2003) Hence, new points of reference were added. Additionally, a **sectoral classification** was introduced; terms like energy security, food security, water security etc. are nowadays commonly applied and understood. Table 1 summarizes some of the broadened security concepts as explained above.

Besides objective criteria for security definitions, the perception of the affected people is of fundamental importance because they will act according to their interpretation of the situation rather than the objective condition. Therefore, A. Wolfers already distinguished in 1962 between objective and subjective security. Objectively, security measures the absence of threats towards acquired values – the actual situation is being measured. On the other hand, the objective security may not necessarily correspond to the security situation as people perceive it. The subjective security therefore means the absence of the fear that acquired values are under threat (Brauch, 2003).

2. Environmental Security

According to Brauch (2005d), environmental security is understood as a so called soft security challenge. Until today no common understanding has been found regarding the definition of environmental and ecological

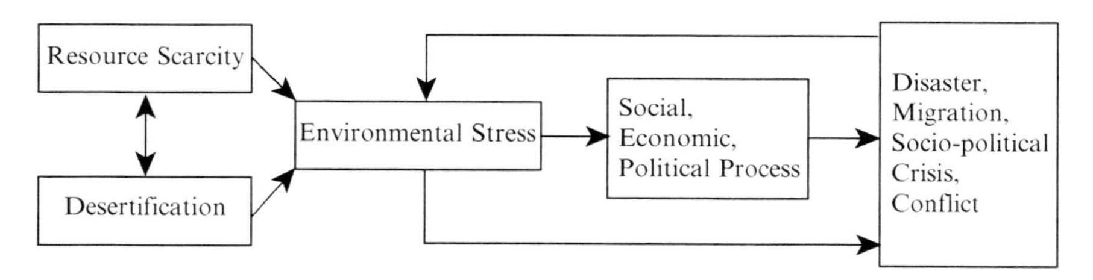

BOX 1. Conceptional relationship between environmental stress and social, economic, political and demographic consequences
Source: Myers (2005), adapted from Brauch (2003)

security. Desertification and Droughts are increasing soft security challenges causing environmental and social vulnerabilities. Under specific global, national, regional and local conditions they may lead to considerable civil consequences, for example, general strikes or revolts caused by starvation (Brauch, 2003).

Especially poor people, who base their livelihood more or less directly on the use of local natural resources, are fundamentally affected by processes decreasing environmental quality and deteriorating the availability of natural resources. Indeed it is a vicious circle because poverty contributes to degradation of natural resources and a degradation/deterioration of natural resources on the other hand again intensifies poverty.

Box 1 depicts the conceptional relationship between environmental stress and social, economic as well as political and demographic consequences. Especially under situations of dense population and direct dependence of people on natural resource use, resource scarcity and desertification both may lead to environmental stress. This again may directly lead to security consequences e.g. due to disasters or migration or have an impact on the social, economic and political situation. If the situation of the people considerably deteriorates, it again may lead to an increase of crises and/or conflict potentials.

3. The Connection between desertification and security

As desertification goes hand in hand with negative environmental changes, there is a linkage between environmental changes and social, economic as well as demographic problems[1]. This linkage has been researched in four phases since the beginning of the 1980s. Much is known about the individual factors leading to drought, migration, crisis and conflicts but the knowledge

[1] See also Kepner and Rubio, (2003)

about the linkages and intertwinements of the disastrous consequences is still incomplete.

Nevertheless, we know that desertification creates a challenge for security and livelihood especially for poor populations. It affects individuals, families, villages, regions and their security in a vicious circle, because poverty contributes to desertification and desertification processes again intensify poverty (Brauch, 2003). The consequences for the population are diverse and may include increases in water scarcity, malnutrition and starvation, migration and deterioration of social status.

As it is one contributing factor for political, social and economic instability, desertification is tightly connected to some of the sectoral security concepts:

- Desertification affects the **food security** because it may cause starvation, which again may lead to migration.
- Desertification affects the **health security** because it may lead to malnutrition and undernourishment leading to increased sensitivity towards diseases and illness; increase child mortality.
- Desertification influences the **livelihood security** because it may force people to leave their homes and migrate as the degraded land can not provide their base of livelihood anymore.
- Desertification has an affect on the **national security** because riots and turmoils due to starvation pose a risk for stability of the countries' regime and the survival of the government.
- Desertification affects **social security** because large scale immigration and scarce resources (water, soil and food) represent a threat for the national identity.
- Desertification concerns the **international security** because cross-boundary mass migration induced by environmental factors may lead to ethnic clashes.

In conclusion, it may be stated that desertification and other environmental damages contribute to economic marginalization of parts of the rural population and therefore are **crisis intensifying factors** in a commonly complex arrangement of numerous conflict and crisis factors.

4. Situation in Central Asia – competition for scarce environmental resources

All Central Asian countries have very limited abundance of the three basic natural resources 'fertile land', 'clean water' and a clean and healthy environment. At least one of those resources, often two, is in short supply in many areas of the region as they are very unevenly distributed.

TABLE 2. Natural geographical and demographic factors

	Kazakhstan	Kyrgyzstan	Tajikistan	Turkmenistan	Uzbekistan
Area	2.717.300 km²	198.500 km²	143.100 km²	488.100 km²	447.400 km²
Population	15 million	5 million	6,5 million	5,5–6 million	25 million
Population growth, p.a. 1980–2000, in %	0,0	1,5	2,2	3,0	2,2
Population growth, p.a. 2000 – 2015, in % (forecast)	0,2	1,1	1,5	1,3	1,3
Geography	≈ 90% steppe	≈ 90% mountains	≈ 90% mountains	≈ 90% desert	≈ 90% desert
Arable land	11%	7%	5%	4%	11%
Rural population per km² arable land	22	236	611	173	342
Freshwater resources per capita/ 1000 m³	7,4	9,5	12,9	11,7	4,6
Freshwater % from other countries	31	0	17	98	86

Source: Grävingholt (2004), page 75 (data from EBRD (2002a); EBRD (2002b, 2002c, 2002d); EBRD (2001a, 2001b); World Bank (2002a); in part own calculations of Grävingholt)

Many different pressures are put upon the already scarce environmental resources and sensitive natural ecosystems of the region. Considerable population growth and hence increase in densely populated areas (e.g. Fergana Valley) combined with inadequate land use activities lead to widespread land degradation, water pollution and general degradation of the ecological situation. Given a high dependency especially of the poor rural population on environmental resource quality and availability, those different pressures create a dangerous mix increasing the potential for (violent) conflicts over land and water resources, migration activities induced by a loss of natural livelihood basis and general food and health crises.

Desert, steppes and high uninhabitable lands are widespread in Central Asia and leave only very limited place for agricultural activities and urbanization. Especially Turkmenistan and Uzbekistan built up large irrigation systems to sustain agricultural production too often based on monoculture, inadequate for the given natural system and conditions and consuming large quantities of water while this resource is scarce, its inflow coming from neighboring countries like Kyrgyzstan and Tajikistan.

According to Grävingholt (2004) Uzbekistan must be considered to be the country with "the largest long-run socioeconomic risk potential" in Central Asia due to its demographic background, its central geographic location and the lack of a plausible development strategy. Nevertheless, Kyrgyzstan and Tajikistan likewise are at risk of encountering strong socioeconomic challenges due partly to their environmental situation, the lowest per capita incomes of the region and because of their direct linkage with developments in Uzbekistans' most densely populated area in the Ferghana Valley.

Kazakhstan has been improving its economic development and management of the economic transformation more successfully mostly due to its wealth in natural resources. Nevertheless, this should not divert the attention from the continued difficult social situation of the large part of the rural population. The transformation of the agricultural sector still remains incomplete, which leads to widespread problems of land degradation especially due to an inadequate legal and institutional settings and performance. Especially small scale farmers often overuse common land areas leading to intense land degradation around villages. With regard to environmental migration, tens of thousands of people in the region of the Aral Sea and other degraded land areas of Central Asia were displaced due to the changes of environmental settings (Myers, 2005).

All in all the question whether those environmental security factors due to desertification problems in the region will lead to an open outbreak of crises and conflicts, depends largely on the setting of other conflict potentials of Central Asia.

According to Grävingholt (2003) one should investigate three different problem areas:

1. Structural conflict causes
2. The ability (or lack of ability) of a society or state to overcome conflicts in a constructive way without applying violence and to find a solution which is legitimate and generally accepted.
3. Conflict aggravating general security risks.

Table 3 summarizes the situation in Central Asia as worked out by Grävingholt (2004). This provides a picture of a high complexity of conflict potentials in the region and shows that environmental changes due to desertification processes are regarded as substantial additional factors potentially weakening the rather sensitive security situation. Nevertheless, the situation differs in several cases between the countries, which should be kept in mind when interpreting the data.

TABLE 3. Crisis and conflict potentials in Central Asia, broken down by problem areas

Structural causes	Constructive conflict-resolution capacity	Conflict-aggravating security risks
• Substantial economic uncertainties persist • Widespread income poverty is a new, difficult experience for many people • Income disparities have grown substantially • Major regional disparities exist within the countries concerned • Social infrastructure (healthcare, education) has deteriorated • A large cohort of young people are without jobs and perspectives • Fertile land is becoming increasingly scarce due to population growth and environmental degradation • Poverty and environmental degradation are causing more and more diseases • National borders cleave grown sociocultural and economic spaces • Natural resources are unevenly distributed among the countries concerned • Certain (regional, ethnic, religious, etc.) identity groups feel systematically discriminated against or are persecuted	• As conflict mediators, government institutions suffer from legitimacy deficits, corruption, and weakness in their implementation capacities • The legal system lacks trust among the population • Civil-society activities are mainly weakly developed and are sometimes massively obstructed • Government intolerance toward and use of force against opposition provokes counterviolence • Government denial of the existence of legitimate conflicts encourages conflict escalation • Western aid regarded as overly regime-oriented (major investments, military, DC) could serve to undercut a conflict moderating role for donors	• The threshold of violence within societies is declining • Government security forces spread uncertainty, not security • Due to the civil war in Tajikistan, many weapons are circulating there as well as in neighbouring regions • Organized crime (trafficking in drugs and arms) is undercutting the government's monopoly on power and encouraging the spread of an illegal economy • Acute crises in one country have the potential to spread quickly to neighboring countries • Dependence on exports of raw materials may lead to unforeseen socioeconomic tensions • Kyrgyzstan and Tajikistan's high external debt is saddling the populations with additional burdens

Source: Grävingholt (2004) page 85

5. Policy Options to prevent desertification induced security problems in Central Asia

Several recommendations can be derived from the existing security hypothesis for policy options with regard to desertification. All of those aim at creating long-term sustainable livelihood conditions and, hence, contribute to prevent conflicts and crises due to negative environmental changes.

The regional planning in rural areas should take into account and harmonize both agricultural production and development of urbanization (development of villages and towns). Furthermore, a strong focus should be placed upon sustainable land use planning and management of desertification prone areas concentrating on vegetation, soil and water conservation strategies and agricultural practices which are in line with the former.

Participation of the local population in planning activities and consideration of traditional knowledge is of utmost importance to achieve long-term sustainable and generally accepted and legitimate results. These should therefore be actively promoted and supported. Legislative, institutional and regulatory measures and performances should be reviewed and adapted to reform land tenure, conservation strategies, harvesting, land use and water management practices and empower local rural communities such that they may actively participate in the recommended changes in land management.

A review and adaptation of federal response plans to open crises (e.g. migration) related to desertification problems should take place such that each country is prepared to manage and overcome desertification related conflicts in a constructive and legitimate way leading to commonly accepted solutions.

An increase of regional cooperation and agreements with regard to transboundary desertification problems (e.g. Aral Sea Crisis; transboundary conflicts over water use) has to be undertaken.

References

Brauch, H. G., 2003, *Desertification – A New Security: Challenge for the Mediterranean?* Presentation at the NATO Mediterranean Dialogue Workshop, Valencia, December 2–5, 2003, (March 27, 2006); http://www.nato.int/science/news/2003/n031202a.htm.

Brauch, H. G., 2005a, *Security Threats, Challenges, Vulnerabilities and Risks – AFES-PRESS Study for UNU-EHS*, Presentation at the Fourth AFES-PRESS Workshop on Reconceptualising Security, "Security Threats, Challenges, Vulnerabilities and Risks", First World International Studies Conference (WISC), Turkey, August 24–27, 2005, (March 27, 2006); http://www.swp.de.

Brauch, H. G., 2005b, *Security and Environment in the Mediterranean and Middle East, Part I,* Presentation at the Integrated Water Resource Management and Security in the Middle

East – A NATO Advanced Study Institute, February 6–17, 2006, Israel, (March 27, 2006); http://www.afes-press.de/html/download_hgb.html.

Brauch, H. G., 2005d, *Reconceptualising Security in the 21st Century. Facing the Challenges of Global Environmental Change and Globalisation*, Opening remarks to Seminar, February 21–23, 2005 (March 27, 2006); http://www.afes-press.de/html/download_hgb.html.

Grävingholt, J., 2003, *Krisenprävention in Zentralasien – welchen Beitrag kann Entwicklungs zusammenarbeit leisten?* in: *Krisen und Konflikte im Osten Europas – Beiträge für die 11. Brühler Tagung junger Osteuropa-Exerten*, Forschungsstelle Osteuropa, Bremen, Auflage Juni, 2.

Grävingholt, J., 2004, *Crisis Potentials and Crisis Prevention in Central Asia: Entry Points for German Development Cooperation,* German Institute for Development Policy, Bonn.

Kepner, W. G., 2003a, *Desertification in the Mediterranean Region. A Security Issue,* Presentation at the NATO Mediterranean Dialogue Workshop, Valencia, December 2–5, 2003, (March 27, 2006); http://www.nato.int/science/news/2003/n031202a.htm.

Kepner, W. G., and Rubio, J. L., 2003, *Executive Summary – NATO-CCMS and Science Committee Workshop on Desertification in the Mediterranean Region - A Security Issue*, December 2–5, 2003, Valencia, (March 27, 2006); http://www.nato.int/science/news/2003/n031202b.htm.

Myers, N., 2005, *Environmental Refugees: An Emergent Security Issue,* Green College, Oxford University, United Kingdom, 13th Economic Forum, May 23–27, 2005, Prague.

CHAPTER 2

FROM GLOBAL ENVIRONMENTAL DISCOURSE TO LOCAL ADAPTATIONS AND RESPONSES: A DESERTIFICATION RESEARCH AGENDA FOR CENTRAL ASIA

LINDSAY C. STRINGER*

Sustainability Research Institute, School of Earth and Environment, University of Leeds, LS2 9JT, UK

Abstract: The UN Convention to Combat Desertification was one of the key conventions emerging from the 1992 Conference on Environment and Development. At this meeting, inter-governmental organisations, NGOs and scientists agreed that desertification and land degradation are 'major economic, social and environmental problems in all regions of the world' and that they 'should be addressed through a United Nations Convention to Combat Desertification' (Chapter 12, Agenda 21; UNCCD, 1994). This chapter outlines the early stages of a novel research project that aims to explore the links between this global environmental discourse on desertification and the practical, local socio-economic impacts and responses to land degradation in the Central Asia region. It first explores the changing role of science in combating desertification. Second, it provides some background to the UN Convention to Combat Desertification, highlighting three key features of its approach: decentralisation, participation and the importance of local knowledge in anti-desertification activities. It then reviews the historical links between the environmental, social, political and economic dimensions of land degradation and desertification in Central Asia. A programme of activities is presented next, which is designed to investigate how the international approaches outlined in the UN Convention to Combat Desertification may impact upon national and local projects and policies to reduce land degradation and improve rural livelihoods. The paper concludes that there is a growing need for detailed, empirical case study research which reflects more intently on the tensions between international and local discourses

*To whom correspondence should be addressed. Lindsay Stringer, Sustainability Research Institute, School of Earth and Environment, University of Leeds, LS2 9JT, UK; e-mail: l.stringer@see.leeds.ac.uk

R. Behnke (ed.), The Socio-Economic Causes and Consequences of Desertification in Central Asia, 13–31.

of desertification. This is paramount in order to better understand the local challenges posed by the desertification issue, as well as contributing towards the creation of more appropriate and effective mitigation strategies at a regional level.

Keywords: Desertification, United Nations Convention to Combat Desertification, Central Asia, Environmental Policy

1. Introduction

In an era of increasing political and popular concern for integrated issues of sustainable development and environmental change, assessing the local impacts of global approaches to dealing with issues of desertification, climate change and biodiversity is of key importance. The year 2006 was designated the '*International Year of Deserts and Desertification*' and as such, this chapter makes a rather timely assessment of the United Nations Convention to Combat Desertification (UNCCD), exploring the potential to apply some of its principles at the local level in the Central Asia region. The layout of the paper is as follows: first, a review of the international desertification discourse and the scientific basis in which it is grounded is presented. Second, focus shifts to look specifically at the Central Asia region, where efforts to combat desertification are taking place in globally unique political, social and economic circumstances. Third, the tensions between the international approach to desertification reduction and the regional context are explored. Finally, the chapter concludes by outlining a preliminary programme of research to explore the relevance of the UNCCD to land users in Central Asia, for whom desertification is part of their daily lives.

Desertification and land degradation are not new environmental problems in Central Asia or the rest of the world. The term 'desertification' itself dates from the 1940s (Warren and Batterbury, 2004), however, the problem only came under the international spotlight for the first time in the 1970s. In 1977, the first global meeting, the UN Conference on Desertification (UNCOD) was convened, to discuss the droughts that had ravaged the Sahel region in the 1960s–70s. During this meeting, 'scientific knowledge' was heavily utilised to justify and legitimate both large-scale and local-level actions to reverse the impacts of degradation and change (Robbins, 2000; Marcussen, 2002). Undeniably, scientific knowledge about desertification has dominated interventions in the past despite the shortcomings associated with its ability to identify and measure degradation and disagreement over what desertification actually is. However, it is only recently that the politicised nature of the use of science has been acknowledged (Forsyth, 2003). While science can help policy develop, it cannot necessarily provide the policy solutions often demanded of it (Eden, 1998). It is only necessary to scratch the surface of topics ranging from genetically modified organisms to climate change to see that science has uncertainties and represents only one way of understanding the world. This is important when considering problems of land degradation and desertification because there are several different opinions on what constitutes degradation, as well as differences in ideas about how it should be identified and measured. Yet, science is powerful and is all too often accepted as fact, while the context from which the science is generated is commonly unacknowledged. This means that scientific datasets can be used to perpetuate generalised (and often inaccurate) statements about issues such as climate change and land degradation (Forsyth, 2003).

For example, publications such as the UNEP World Atlas of Desertification (1992) have used scientific information to reinforce the discourse of a global desertification crisis, with claims that up to 70% of the agricultural land in the world's drylands is affected by land degradation and that one sixth of the world's population is under threat from desertification (Dregne *et al.*, 1991; Agnew and Warren, 1996). In many regions, land use and vegetation cover were reported to have altered over time, and soils were eroding at accelerated rates leading to a reduced capacity of these areas to support people, livestock and wildlife (Reynolds and Stafford-Smith, 2002). Coupled with population growth and climate change uncertainties, sustainable natural resource utilisation and rural livelihoods were considered to be under increasing threat in many of the world's drylands (Brookfield, 1995). However, this popularised discourse did not differentiate between environmental change and environmental degradation. It glossed over the differences and presented the desertification issue as being in urgent need of attention, with dire social and environmental consequences if action was not taken (Roe, 1991; Sutton, 1999).

Thirty years on from the UNCOD meeting and fourteen years on from the production of the World Atlas of Desertification (UNEP, 1992), problems of land degradation and desertification, and their links with both unsustainable development and global poverty, remain strong. This is despite several inter-governmental, practical and awareness-raising initiatives to reduce the problem, including the designation of 2006 as the '*International Year of Deserts and Desertification*' and a number of locally managed capacity-building events. However, to say that the (primarily science-based and technical) policies and interventions of the last three decades have been futile would be both unhelpful and incorrect, as approaches to understanding land degradation and desertification have evolved considerably during this time.

Extensive debate has ensued over the magnitude, severity, causes and effects of observed environmental changes and the scientific grounding on which policy responses to perceived land degradation are based (Dahlberg, 2000). This is because interpretations of environmental changes vary considerably depending on local perspectives, different scientific approaches, national policy environments and local political agendas (e.g. Leach and Mearns, 1996). In light of this, scientists have started to look beyond the biophysical manifestations of degradation towards their socio-political and economic drivers, as investigations have become broader and multifarious; reaching past the biophysical processes of degradation to explore the social, economic, environmental, political and historical facets of environmental change. As a result, perspectives that challenge the belief that the environment can be managed and governed exclusively through the application of scientific principles have become more widely acknowledged, in both research and policy spheres (Bäckstrand, 2004).

Notably, at the international level, local knowledges have been afforded more respect and credibility, and are beginning to be valued rather than seen as backward and inferior to science. Agreements such as the United Nations Convention to Combat Desertification (1994) present local knowledge as a valuable resource that should be integrated with scientific knowledge, to define, monitor and measure desertification, as well as contribute to effective and appropriate conservation policy and practice. These gradual changes to the ways in which desertification is both identified and addressed mean that (in theory) the people who live with land degradation should become more empowered to deal with the problem.

2. The UNCCD (1994)

The idea of a Convention to Combat Desertification has long been controversial. It first stemmed from the African nations, prior to the 1992 United Nations Conference on Environment and Development (Chasek and Corell, 2002). These states considered that problems of poverty, drought, inequality and food insecurity were impeding their sustainable development (Warren, 2002), but were being largely sidelined by the international political community. International focus had instead been on climate change and biodiversity issues (Toulmin, 2001); issues that were considered to be of most concern to countries of the developed world. A number of developed states (particularly those in the European Union), initially opposed the idea of a convention (Chasek and Corell, 2002), despite both moral and poverty dimensions to the problem. They considered desertification not to be a truly global issue (Thomas and Middleton, 1994; Stiles, 1995) and argued that biophysical environmental changes are manifest at the local level, largely as a cumulative effect of individual farmer actions and decisions (Batterbury et al., 2002).

After considerable debate, an agreement of sorts was reached. It was acknowledged that although the effects of land degradation may only become apparent at the local level, links with global processes could not be ignored. Large-scale environmental conditions (e.g. climate, soil and hydrological patterns) which influence processes of degradation and change are defined at national, regional and global scales. These can drive problems of drought, climate change and biodiversity loss, while socio-political and economic factors such as markets, technological changes and human migration also have international dimensions. Therefore, in response to the African request and in conjunction with documentation in Chapter 12 of Agenda 21, in its 47th Session in 1992, the UN General Assembly resolved to establish the Intergovernmental Negotiating Committee on the Desertification Convention. Following negotiations, the Convention to Combat Desertification was adopted in Paris on 17 June 1994 and opened for signature on 14–15 October 1994 (Toulmin, 1995). It entered into force on 26 December 1996, 90 days after the fiftieth ratification was received. By January 2006 it had gained the signature and support of 191 countries.

Nevertheless, the debate on the scale issue did not end with the agreement to produce an international convention, as the UNCCD acknowledges that solutions have to be found at the local, national, sub-regional and regional levels (Kjellen, 2003). This task is facilitated through the inclusion of regional implementation annexes in the UNCCD and an emphasis is placed on National Action Programmes (NAPs) and Sub-Regional Action Programmes (SRAPs) as mechanisms of implementation. Such moves away from the international level were intended to allow activities to combat desertification to be more context-specific; politically, environmentally, socially and economically.

A further achievement of the UNCCD is that it has firmly embedded the land degradation and desertification issue in broader debates. These concern issues such as: access to and control over natural resources; participation; democratisation; synergy with other problems such as biodiversity loss and climate change, and sustainable development. It recognises the need for the decentralisation of natural resource management through broad-based local knowledge inputs into decision-making and policy processes (e.g. see Agrawal and Gibson, 1999; Kiker et al., 2001; Ribot 2004; Larson and Ribot, 2004, Stringer and Reed 2007). However, determining how science and local knowledges should be integrated into anti-desertification policy is not straightforward and involves the development of flexible ways in which scientists, local actors, and their knowledges can interact in a democratic way. According to Forsyth (2003), one way of doing this is to use local knowledge as a starting point in research on environmental change and to use 'science' as a means to extend these to wider areas of management.

Recent trends have also seen the rise of participatory approaches such as 'adaptive co-management', in which 'experts' from a range of disciplines and 'lay' people learn together about a particular situation and develop joint solutions (Holling, 1978; Daniels and Walker,1996; Kiker et al., 2001). However, whether these approaches are appropriate to Central Asian societies where people have been long-estranged from participation, has received very little critical attention.

In summary, while science has long played a key role in identifying, monitoring and mitigating desertification, it is only recently that international discourses have acknowledged the potentially valuable contribution of local knowledges, through the promotion of decentralised, participatory activities and strategies to address the problem. Whether (and how best) scientific and local knowledges can be effectively brought together remains a complex question (Stringer and Reed, 2007), particularly in regions with little history of public involvement in environmental decision making.

3. The Central Asia context: A review

This section moves to specifically consider the problems faced by the Central Asian region by exploring the historical environmental, political, social and economic aspects of society-environment relationships. It considers why all the Central Asian countries are thought to be either affected or severely affected by problems of desertification (UNCCD website, March 2006) and presents the historical legacy upon which contemporary national policy decisions are being built.

3.1. FROM THE PAST TO THE PRESENT

In terms of the environment, the Central Asian countries of the Former Soviet Union, like many other regions of the world, are naturally disposed to problems of land degradation and desertification. The combination of a harsh climate, steeply sloping land (in places) together with an unevenly distributed resource potential has historically determined the patterns of human settlement and resource use. Until the Soviet era, many people were living within and coping with these natural environmental limits. They were managing the inherent environmental heterogeneity through practices such as (semi-) nomadic pastoralism, which allowed the flexible continuation of agricultural activities throughout the year (Kerven et al., 2004). In Kazakhstan for example, pastoralists kept mixed herds of horses, goats, sheep and camels. They migrated annually across thousands of kilometres to take advantage of seasonal spatial environmental fluctuations. This permitted household resources to be used to their full potential (Kerven et al., 2004). Although by the mid-19th century migratory movement was already declining, this represented a flexible natural resource use strategy that recognised environmental complexity and unpredictability (Robinson et al., 2003).

As outlined by Cherp and Mnatsakanian (2003), at this time the institution of private property for land did not extend beyond a very narrow circle of noble landowners in European Russia and was even less developed in more traditional societies of Siberia and Central Asia. For the majority of the population, dependence was placed on a freely accessible natural resource base for day-to-day food, firewood and construction[1]. Indeed, until the early 20th century, most of the region's productive land was either in village communal property or in large estates (Cherp and Mnatsakanian, 2003). However, with the creation of the Soviet Union in the 20th Century and the implementation of large-scale initiatives such as the 'virgin lands campaign' in Kazakhstan, and huge irrigation projects to meet cotton production targets in the Aral Sea region (e.g. Saiko,1995), important environmental, economic and social changes took place. Patterns of natural resource use were comprehensively reorganised, as agricultural production was restructured, primarily around state farms (*sovkhozy*) and collective farms (*kolkhozy*). For example, in Kazakhstan in the 1930s, Stalin first imposed collectivisation and involuntary settlement on pastoralist communities, then later (after political and economic control had been established) adopted a policy of centrally-directed livestock migration (Kerven et al., 2004). At the same time, arable land throughout the republics was earmarked by the state for specific production goals. For example, countries such as Uzbekistan, Turkmenistan and Tajikistan saw specialisation in cotton production, Kyrgyzstan was the main source of the region's wool production and Kazakhstan became the key producer of grain, primarily as a result of the 'virgin lands' campaign (Spoor, 1999).

Besides being agricultural production units, the *sovkhozy* and *kolkhozy* structures also provided important social functions, including basic income and food to members, as well as social and health services in rural areas (Spoor, 1999). Productivity decisions regarding the crops grown, sown area, livestock numbers and so on were made directly by the leadership of the Communist Party of the Soviet Union, the republic, the region and the dis-

---

[1] Care must be taken not to over-romanticise the past, as it would be incorrect to portray pre-Soviet central Asia as an unspoiled 'Eden'. However, the point I wish to make is that local people were able to make their own decisions within the structures of prevailing social and cultural norms, often choosing to acknowledge environmental heterogeneity and use it to their advantage.

trict, with the government both providing the necessary inputs and procuring the outputs (Suleimenov and Oram, 2000). The system that developed represented a top-down and highly centralised mode of control, which prevailed in all spheres of local life. Public opinion was largely disregarded and the resulting legacy was popularised as one of widespread environmental degradation[2] (Cherp and Mnatsakanian, 2003).

Since gaining independence in the early 1990s, the countries of the Central Asia region have experienced varying degrees of socio-economic and political reforms, including shifts towards democratisation, decentralisation, privatisation, improved access to information for ordinary citizens and land reforms (UNCCD website, March 2006). Despite some similarities in the restructuring of previously dominant *sovkhozy* and *kolkhozy* across the region, the ways in which the reforms have played out has varied between countries. At the macro level, the economic shifts taking place have not been uniform throughout the region and significant differences between policies in different countries have emerged. For example, Kazakhstan has tended towards a strategy of economic liberalisation, while Uzbekistan has been more gradual in its sequencing of reforms (Alam and Banerji, 2000). Despite these differences, each of the changes has had both direct and indirect implications for environmental protection and is likely to have had profound impacts on local abilities to assess and respond to environmental changes including land degradation and desertification. However, empirical information on what the limitations and opportunities are, is notably absent from the academic literature.

In summary, the creation and break up of the Former Soviet Union left an unequal distribution of environmental and social problems across the independent states. What remained was an imprint that not only shaped the contemporary social, political, economic and environmental landscape but a legacy that also provided the foundations for nation-building in the newly independent states. This in turn has an important impact on the environmental problems and policies we see in place today, as well as on local inclinations and capacities to recognise and respond to environmental changes. The next section explores some of these problems looking in particular at those linking to desertification and its mitigation.

[2] Whilst environmental degradation in the Soviet era has received most attention here, it is important to note that the outcomes of socialist rule were not all environmentally negative. For example, the Soviet Union pioneered a unique network of nature protection areas (known as *zapovedniks*). These territories were strictly protected and today hold a potentially useful role in assessing long-term environmental changes, since their ecosystems remain largely undisturbed (Cherp and Mnatsakanian, 2003).

4. Desertification in Central Asia

In the 1990s, while the international community was focusing on the need for and articulation of a Convention to Combat Desertification, environmental problems including land degradation and desertification were playing an important role in triggering the social, political and economic changes outlined previously (Cherp & Mnatsakanian, 2003). Agricultural productivity had been declining and the large state and collective farms were suffering from the effects of low yields, technological stagnation and a general inefficiency of resource use (Spoor, 1999). The natural resource potential of many areas was historically known to be limited (primarily due to climatic conditions) yet the "command and control" approach of the centralised Soviet administration had failed to acknowledge the limits to environmental predictability.

While planning had relied on scientific expertise and the region's scientists had received comprehensive scientific training, the focus of research on agricultural productivity and meeting centrally-developed targets meant that much of the data and knowledge focused primarily on the state of affairs at the macro-level. Furthermore, no single methodology was used to monitor and measure land degradation, nor were there any standardised indicators (Robinson et al., 2003). This meant that assessment of the land's condition in the Former Soviet Union was not necessarily comparable between different parts of the region. Robinson et al., (2003) suggest that one of the main pieces of evidence used to support estimates of the extent and location of land degradation is a map of degradation intensity in Central Asia that was compiled at the Institute of Deserts in Turkmenistan (Babaev, 1985). However, this has been widely criticised because the methodologies for the classification of different degrees of degradation are not provided, even in the papers which accompany the map. Also, since many of the estimates of degradation were based on these types of maps and did not always involve fieldwork, it meant that local environmental change and diversity was (and to an extent, still is) relatively understudied. Therefore, whether the international discourse of an impending desertification crisis holds true at the local level is unknown.

While all the countries of the Former Soviet Union have ratified the UNCCD in an attempt to stem land degradation and desertification (see Table 1), the dynamics of the problem also remain unknown, particularly since the shift towards new market economies has allowed rural people to have more control over the decisions affecting their lives and their environments.

Despite the problems with non-uniform assessment methodologies, the impacts the historical legacy has had on processes of desertification and

TABLE 1. Central Asian ratification of the UNCCD

Country	Date of signature	Date of Ratification (R), Accession (AC) or Acceptance (AT)	Date of Entry into Force	Date of NAP
Kazakhstan	14/10/1994	09/07/1997 (R)	07/10/1997	1997, 2001, 2005
Kyrgyzstan	—	19/09/1997 (AC)	18/12/1997	2000
Tajikistan	—	16/07/1997 (AC)	4/10/1997	2001
Turkmenistan	27/03/1995	18/09/1996 (R)	26/12/1996	1997
Uzbekistan	07/12/1994	31/10 1995 (R)	26/12/1996	1999

land degradation are considered to have manifest themselves in various different ways throughout the region. Table 2 presents government views of the problems in need of most urgent attention in three countries in the region, as documented in their NAPs. Perceived problems include the salinisation of many (inappropriately) irrigated soils throughout the region, which is thought to have had negative economic impacts on agricultural yields. Water supplies too are largely saline in much of the region and this has caused the overgrazing of areas close to non-saline watering holes, while military manoeuvres and weapons testing in the Soviet era are thought to have caused the degradation and contamination of large parts of the region. Another problem is that of wood cutting for use fuel. This is said to have caused a loss of vegetation, leaving sandy soils more vulnerable to wind and water erosion. In some cases, it has led to the formation of dunes which have caused problems for gas and oil extraction infrastructure, as well as for local people. This also has important economic implications, particularly because one of the main reasons for macroeconomic improvement in the region since independence is due to the extraction and export of natural mineral and energy resources (Raissova and Sartbayeva-Peleo, 2004). Nevertheless, as indicated by the dated reference lists at the end of each NAP, many of what are considered to be the most urgent land degradation and desertification problems are based on data from the Soviet period and lack contemporary empirical evidence.

The nature of these problems illustrate the strong interdependence of the social, economic and environmental elements of development, and highlight the need to move away from short-term planning towards a focus on longer-term sustainability issues. This will require not only the incorporation of environmental concerns into economic and social policies but also the building of new institutions, political will, the creation (and enforcement) of new legislation and also the building of both state and local capacities to monitor and manage land degradation and desertification. The first step however,

TABLE 2. Desertification issues in Central Asia (based on information supplied in each country's National Action Programme to Combat Desertification: Kazakhstan NAP, 1997; Turkmenistan NAP, 1997; Uzbekistan NAP, 1999)

	Kazakhstan	Turkmenistan	Uzbekistan
Climate and water	Dry continental climate. Water shortages are common due to over consumption and over-engineering of the river system in the past	Irrigation is necessary for much of the land and 92% of water is used in agricultural activity.	Wind speeds are not particularly high but dust storms occur up to 64 days/yr. Drought and climate characteristics in dry areas & over-extraction of water for irrigation have led to an imbalance in water supply and demand. 90.4% of water used is for irrigation. This increased sharply from 1975–1985, when 1 million ha of 'new' land was irrigated.
Soils	42.1% of the land area is occupied by salinised and solonetz soils	Mostly gypsiferous grey-brown soils and takyr/takyr-like soils.	Soils include: flood-plain alluvial, lerovo-greyish, desert sandy, desert takir, greyish-brown deserts, irrigated greyish, light greyish, usual greyish, mountain brown and high mountain soils.
Key causes of land degradation and desertification	Massive-scale ploughing of virgin lands in 1950s & 60s; overgrazing of rangeland areas due to increase in ploughed land and rising livestock numbers; nuclear testing and mining has rendered much land unsuitable for human habitation.	Reduction of forest area; inappropriate irrigation in combination with saline waters, poor drainage and a shallow water table; concentration of livestock near suitable water sources.	Military training, manoeuvres of overland armies, nuclear and strategic weapons testing in the Soviet era; inappropriate redirection of water courses leading to changes to the water balance and salinity; felling of wood and shrubs for fuel (including root-cutting); increases in cattle numbers; and widespread mono-cropping of cereals and no crop rotation.

(continued)

TABLE 2. (continued)

	Kazakhstan	Turkmenistan	Uzbekistan
Manifestations of desertification	Sand and dust storms have become commonplace. Vegetation cover is scarce and expanses of sand and dry climate are expanding. Biodiversity loss has ensued. Forest degradation in Rudny and South Altai, and in the Charadarinsk region in the south, almost a third of turanga forests changed to chingil bushwood. Areas close to villages, cattle pens and wells suffer most severely from livestock-induced degradation.	From 1983–1988 the forested area in Turkmenistan declined by 994,000ha, causing a reduction in the productivity of the upper soil layers and formation of dunes. Agricultural yields are affected by salinity. Cotton is the main crop and this requires irrigation, leading to further salinity. Much of the water is very salty (more than 15 g salt per litre). Vegetation is only maintained throughout the spring and summer leading to the overgrazing of rangeland during the rest of the year.	Overgrazing has caused greater quantities of mobile sands which act as centres for deflation. Wind and water erosion is common as upper soil horizons have been damaged and vegetation cover and biodiversity have been lost. 73% of all agricultural areas are subject to wind erosion and 18% are subject to water erosion. Inappropriate agricultural techniques and irrigation of arable areas have caused increased soil salinity and dust storms are frequent in the Aral region.
Economic impacts	Reduced harvesting capacity & decreased livestock productivity and breeding potential, have led to the decreased viability of exports from the agricultural sector & slow development of the food industry.	Agricultural yields declined due to salinisation & livestock productivity declined due to overgrazing (especially near water holes). Oil and gas extraction has taken place over the past 20 years and moving sand dunes have caused problems for oil and gas infrastructure (pumping stations, labour settlements etc).	Dust and salt movement limit the productivity of agricultural lands, as does the accumulation of pesticides and herbicides in the soil. 46.8% of irrigated lands are saline and of this, 6.6% are strongly saline.

Social impacts	Annually, hundreds of thousands of people are migrating away from desertified areas. Primarily this is because desertification has affected their livelihood sustainability & health. There is a lack of water uncontaminated by dust storms and suitable for human consumption. Public funding for education, healthcare and poverty reduction has sharply decreased.	Rural-urban migration and a lack of social services have had an important socio-economic impact.	The agricultural sector is oversaturated with labour. One third of public sector workers are engaged in agriculture. The gap between rich and poor is increasing, despite a growth of real incomes of 8.8% since 1996. Migration from rural-urban areas is causing labour surpluses in the towns. Social service provision (medical care, education) are lacking in the countryside too.

before more appropriate policies and planning processes can be developed, is to collect new data to explore the extent and magnitude of the problems under current conditions. Without knowing the degree and scale of the issue, it is difficult to see how appropriate and effective policies can ensue.

The UNCCD provides an important starting point for many of these changes to take place because it moves beyond the prescriptive set of technical means to identify and address desertification that early attempts to combat desertification favoured. Instead, it provides a potentially useful framework to maintain the position of desertification within global political and donor arenas, sustaining global awareness of the desertification issue in the overall framework of sustainable development. Despite these opportunities however, to date there has been little reflection on the suitability of the UNCCD's decentralised, participatory approaches to natural resource management, for the specific socio-political circumstances of the transition states. It is unclear whether the historical environmentally sensitive forms of land management have been lost from local knowledge bases, or whether the serious economic difficulties experienced over the past 16 years have led to new ways of thinking about people's relationships with their environments. Little is known about the micro-level impacts of national social, economic and political transition on the ways in which land degradation, drought and desertification are defined and addressed. Also, it is unclear whether such decentralised approaches can be easily implemented in a society where public participation has been historically absent. These questions form some of the key gaps in desertification studies and are in urgent need of improved understanding. Furthermore, information in the Western scientific literature on the impacts of the socialist system on the environment and local societies is largely absent, and remains narrowly confined to Russian-language publications (Robinson et al., 2003).

One possible way of addressing some of these challenges, is outlined in the next section. The research agenda that is presented therein draws attention to some of the challenges the Central Asia region faces in moving between international discourse and positive local environmental outcomes.

5. Global desertification discourse and local adaptations and responses: Taking desertification research forward in the Central Asia region

An important first step towards understanding desertification in Central Asia is to evaluate the appropriateness of the environmental and social approaches of the UN Convention to Combat Desertification, exploring how these may impact upon national and local projects and policies to reduce land degradation and improve rural livelihoods. This would not only provide a useful case study for the Central Asia region but would also allow the exploration of wider epistemic and methodological questions relating to the drawing together of scientific and indigenous knowledges and the implications for such approaches to wider processes of environmental governance. Funding is currently being sought by the author to undertake such work in two or three small rural settlements in one case study country.

Five key research questions will guide the proposed research:

1. How have land degradation and desertification been identified scientifically within the study region and how are they presented a) within national policy (particularly in National Action Programmes to Combat Desertification) and b) within international environmental discourse?
2. How do the people who make a living from the land manage and evaluate desertification and land degradation?
3. How have recent economic, political and social changes affected local capabilities to address the problems identified in 2)?
4. How do these representations of land degradation vary across local, national and international levels, and what are the reasons behind elements of agreement and difference?
5. What are the environmental and social implications of these similarities and differences?

Kazakhstan has been selected as an initial area of focus. However, it is hoped that the research will be able to incorporate a greater number of case study countries as it proceeds.

5.1. PROPOSED METHODS

Research situated at the interface of environment and society is methodologically demanding due to the diversity of ecological, cultural and socio-economic factors that need to be considered (McKendrick, 1999). While the proposed research agenda can be located broadly within the growing body of work following the political ecology tradition, it is important not to be methodologically or theoretically bound by narrowly defined academic disciplinary boundaries. As such, an interdisciplinary approach is needed, which draws on theories and methods from a range of disciplines, including geography, agricultural sciences, politics, and development studies. This allows a more detailed and comprehensive understanding of the perspectives of different stakeholder groups, while permitting both social and environmental facets of the same research question to be addressed. It also permits investigations to move between different scales to compare and contrast international discourses and local environmental outcomes, culminating in an in-depth exploration of the challenges the Central Asia region faces in addressing the desertification issue.

To address question 1, a desk-based review of the theoretical literature, the political, environmental and economic characteristics of the case study region will be carried out. Anti-desertification and sustainable development policies of the study country will also be analysed using methods of discourse analysis. This will assist in highlighting the main desertification narratives or 'stories' and will provide important information with which the field data can be triangulated. The remaining research questions require the collection of primary data, using techniques such as: household questionnaire surveys, interviews with scientists, policymakers and NGOs and participatory rural appraisal techniques (transect walks, seasonal calendars etc.) with rural people. Use of multiple methods will allow the triangulation of the results and where contradictions emerge, the iterative nature of data collection will aid broader and deeper explanation (Stringer and Reed, 2007). Fieldwork is planned for a period of 6–8 weeks (depending on the timing and availability of funding).

It is envisaged that the outputs from this research will be useful and interesting to a number of different stakeholders: 1) those involved with UNCCD implementation at the international, regional, sub-regional and national levels 2) NGOs 3) academics and the scientific community. Following completion, a summary report from the project will be circulated to relevant stakeholders and a minimum of two journal articles will be submitted to high impact, international, peer-reviewed journals. Ideally, some of the proposed outputs will also be translated into Russian to allow wider dissemination of the findings.

6. Conclusion

This chapter has explored the current international approaches to addressing desertification and has situated these within a general review of environment and society relations in Central Asia. A number of challenges and gaps in our understanding about desertification have been highlighted, including a growing need for detailed local-level empirical research. A research agenda has been presented which hopes to address some of the identified gaps, and when executed, will provide long over-due reflection on the tensions and opportunities associated with international and local discourses of desertification.

Despite the socio-economic and environmental challenges that the Central Asia region has faced over the past 16 or so years, it is important to conclude by considering the positive changes that have taken place. Transition has brought with it many opportunities for economic, social and environmental change and it is important that the positive legacies from the past can be built upon. Similarly, the negative legacies can serve as wider lessons to other regions in order to avoid new threats and reverse the environmental degradation of the past (Cherp and Mnatsakanian, 2003). Of key importance however is the continuation of research into the causes and effects of desertification, its wider socio-economic impacts and its links with broader sustainable development issues. Given the close relationship between the vulnerability of dryland populations, livelihood sustainability, desertification, poverty and climate change, the links and synergies with broader scale development processes can only strengthen, as attempts continue to decrease the impacts of desertification, land degradation and drought throughout the world.

References

Agnew, C., and Warren, A., 1996, A framework for tackling drought and land degradation, *Journal of Arid Environments,* **33**:309–320.

Agrawal, A., and Gibson, C. C., 1999, Enchantment and disenchantment: the role of community in natural resource conservation, *World Development,* **27**:629–49.

Alam, A., and Banerji, A., 2000, Uzbekistan and Kazakhstan: a tale of two transition paths?, *World Bank Policy Research Working Paper,* **2472**.

Babaev, A. G., (Ed.) 1985, Map of anthropogenic desertification of arid zones of the USSR, Ashgabat, Turkmenistan: Institute of Deserts.

Bäckstrand, K., 2004, Science, Uncertainty and Participation, *Environmental Politics* **13**(3):650–656.

Batterbury, S. P. J., Behnke, R. H., Doll, P. M., Ellis, J. E., Harou, P. A., Lynam, T. J. P., Mtimet, A., Nicholson, S. E., Obando, J. A., and Thornes, J. B., 2002, Responding to desertification at the national scale, in: *Global Desertification: Do Humans Cause Deserts?,* J. F. Reynolds, and D. M. Stafford-Smith, ed., Dahlem University Press, Berlin.

Brookfield, H., 1995, Postscript: the 'population-environment nexus' and PLEC, *Global Environmental Change* 5, pp. 381–393.

Chasek, P. S., and Corell, E., 2002, Addressing desertification at the international level, in: *Global Desertification: Do Humans Cause Deserts?* J. F. Reynolds, and D. M. S. Stafford-Smith, ed., Dahlem University Press, Berlin.

Cherp, A., and Mnatsakanian, R., 2003, Environmental degradation in Eastern Europe, Caucasus and Central Asia: past roots, present transition and future hopes, Central European University, Budapest.

Dahlberg, A. C., 2000, Interpretations of environmental change and diversity: a critical approach to indications of degradation—the case of Kalakamate, northeast Botswana, *Land Degradation & Development,* **11**(6):549–562.

Daniels, S. E., and Walker, G. B., 1996, Collaborative learning: improving public deliberation in ecosystem-based management, *Environmental Impact Assessment Review,* **16**:71–102.

Dregne, H., Kassas, M., and Rozanov, B., 1991, Status of desertification and implementation of the United Nations Plan to Combat Desertification, *Desertification Control Bulletin,* **20**(6).

Eden, S., 1998, Environmental issues: knowledge, uncertainty and the environment, *Progress in Human Geography* **22**:425–432.

Forsyth, T., 2003, *Critical Political Ecology,* Routledge, London.

Holling, C. S., 1978, *Adaptive environmental assessment and management*, Wiley, New York, USA.

Kazakhstan NAP, 1997, Kazakhstan National Action Programme for Combating Desertification and Mitigating the Effects of Drought.
Kerven, C., Alimaev, I. I., Behnke, R., Davidson, G., Franchois, L., Malmakov, N., Mathijs, E., Smailov, A., Temirbekov, S., and Wright, I., 2004, Retraction and expansion of flock mobility in central Asia: costs and consequences, *African Journal of Range and Forage Science*, **21**(3):91–102.
Kiker, C. F., Milon, J. W., and Hodges, A. W., 2001, Adaptive learning for science-based policy: the Everglades restoration, *Ecological Economics*, **37**:403–416.
Kjellen, B., 2003, The saga of the Convention to Combat Desertification: The Rio/Johannesburg process and the global responsibility for the drylands. *RECEIL*, **12**:127–132.
Kyrgyzstan NAP, 2000, Kyrgyzstan National Action Programme for Combating Desertification and Mitigating the Effects of Drought.
Larson, A. M., and Ribot, J. C., 2004, Democratic decentralisation through a natural resource lens: an introduction, *European Journal of Development Research*, **16**:1–25.
Leach, M., and Mearns, R., 1996, *The Lie of the Land*, The International African Institute in association with James Currey, Heinemann, Oxford.
Marcussen, H. S., 2002, International conventions and the environment: what future for the international Convention to Combat Desertification?, in: Proceedings of the CCD Workshop, 26–27 February 2002, H. S. Marcussen, I. Nygaard, and A. Reenberg, eds., Denmark.
McKendrick, J. H., 1999, Multi-method research: an introduction to its application in population geography., *Professional Geographer*, **51**:40–50.
Raissova, A., and Sartbayeva-Peleo, A., 2004, From Rio to Johannesburg: Comparing Sustainable Development in Kazakhstan, Uzbekistan, and the Kyrgyz Republic, in: In the Tracks of Tamerlane: Central Asia's Path to the 21st Century, D. L. Burghart, and T. Sabonis-Helf, ed., pp. 245–258.
Reynolds, J. F., and Stafford-Smith, D. M., 2002, Do humans cause deserts? in: "Global Desertification: Do Humans Cause Deserts?, J. F. Reynolds, and D. M. S. Stafford-Smith, ed., Dahlem University Press, Berlin.
Ribot, J., 2004, Waiting for Democracy: Politics of Choice in Natural Resource Decentralization, World Resources Institute.
Robbins, P., 2000, The practical politics of knowing: state environmental knowledge and local political economy, *Economic Geography* **76**:126–144.
Robinson, S., Milner-Gulland, E. J., Alimaev, A., 2003, Rangeland degradation in Kazakhstan during the Soviet era: re-examining the evidence, *Journal of Arid Environments*, **53**:419–439.
Roe, E. M., 1991, Development narratives, or making the best of blueprint development, *World Development*, **19:**287–300.
Saiko, T. A., 1995, Implications of the disintegration of the former Soviet Union for desertification control, *Environmental Monitoring and Assessment*, **37**:289–302.
Spoor, M., 1999, Agrarian transition in former Soviet central Asia: a comparative study of Kazakhstan, Kyrgyzstan and Uzbekistan, Working paper 298, Institute of Social Studies, The Hague.
Stiles, D., 1995, *Social aspects of sustainable dryland management*, Wiley, Chichester.
Stringer, L. C., and Reed, M. S., 2007. Land degradation assessment in southern Africa: integrating local and scientific knowledge bases. *Land Degradation and Development* **18**:99–116.
Sutton, R., 1999, *The Policy Process: an Overview*, Overseas Development Institute, London.
Suleimenov, M., and Oram, P., 2000, Trends in feed, livestock production and rangelands during the transition period in three central Asian countries, *Food Policy*, **25**:681–700.
Tajikistan NAP, 2001, Tajikistan National Action Programme for Combating Desertification and Mitigating the Effects of Drought.
Thomas, D. S. G., and Middleton, N. J., 1994, *Desertification: Exploding the Myth*, Wiley, Chichester.

Toulmin, C., 1995, Combating desertification by conventional means, *Global Environmental Change,* **5**:455–457.
Turkmenistan NAP, 1997, Turkmenistan National Action Programme for Combating Desertification and Mitigating the Effects of Drought.
UNEP, 1992, *World Atlas of Desertification*, Edward Arnold, London.
United Nations Convention to Combat Desertification, 1994, United Nations Convention to Combat Desertification in Those Countries Experiencing Serious Drought and/or Desertification Particularly in Africa: Text with Annexes UNEP, Nairobi.
Uzbekistan NAP, 1999, Uzbekistan National Action Programme for Combating Desertification and Mitigating the Effects of Drought.
Warren, A., 2002, Land degradation is contextual, *Land Degradation and Development,* **13**:449–459.
Warren, A., and Batterbury, S., 2004, Desertification, in: The Routledge Encyclopaedia of International Development. T. Forsyth, ed., Routledge, London Accessed on 25 August 2005; http://www.u.arizona.edu/~batterbu/papers/desertificationforsythweb.htm

CHAPTER 3

CAUSES AND SOCIO-ECONOMIC CONSEQUENCES OF DESERTIFICATION IN CENTRAL ASIA

CAUSES AND CONSEQUENCES OF DESERTIFICATION

LAPAS ALIBEKOV*[1] AND DAVLAT ALIBEKOV[2]

[1]*Department of Physical Geography and Geoecology, Samarkand State University, 703004, 15 University Blvd. Samarkand, Uzbekistan*
[2]*Department of Economics, Samarkand State University, 703004, 15 University Blvd. Samarkand, Uzbekistan*

Abstract: Agriculture in Central Asia was mismanaged for decades under a centralized command economy. The continuing legacy of this period is severe land degradation in the form of soil salinization and erosion, elevated groundwater levels caused by poorly managed irrigation systems, the drying of the Aral sea, and the chemical and nuclear pollution of water and soil. This paper provides an overview of how these failures occurred and describes some of their negative social and economic impacts.

Keywords: land degradation, salinization, Central Asia, Uzbekistan, irrigation

1. The causes of desertification in Central Asia

Widespread desertification occurs in many parts of Central Asia. This environmental degradation is caused by socio-economic factors, the main reason being the overall incompatibility between regional production systems, which were formed in the past, and the regional ecosystem. This mismatch

---

*To whom correspondence should be addressed. L. Alibekov, Department of Physical Geography and Geoecology, Samarkand State University, 703004, 15 University Blvd. Samarkand, Uzbekistan; e-mail: davlat1982@yahoo.com

R. Behnke (ed.), The Socio-Economic Causes and Consequences of Desertification in Central Asia, 33–41.

has occurred as a result of ignoring both the laws of nature and economics, as evidenced by the wasteful exploitation of water, land and other natural resources under the centralized economic and administrative system. This overexploitation has taken several forms.

The monoculture of cotton destabilized both the environment and the regional agricultural system by constraining the growth of other agricultural subsectors such as horticulture and grain, forage and livestock production. Monoculture and the expansion of irrigation also lead to shortages of irrigation water, the drying of the Aral Sea, and eventually to the Aral ecological crisis and the spread of desertification processes across a large territory. The diversion of water for irrigation needs from Amu Darya and Syr Darya rivers disturbed the balance between the flow of water into the Sea and the loss of water to evaporation, which has been disastrous not only for the sea itself, but for the whole Aral basin, leading to large scale and potentially irreversible environmental transformations.

An additional important factor leading to desertification in Central Asia has been inappropriate irrigation which has lead to salinization. Large-scale dams were built starting in the 1960s. These have promoted salinization and increased ground water levels in low lying areas characterized by poor natural drainage. The expansion of areas of irrigated land at higher elevations has also damaged the productivity of older irrigated areas in the valleys of rivers or on the margins of plains adjacent to mountains, which are being waterlogged by underground flows from higher areas of newly developed cultivation. A clear example of this process is the declining productivity of older irrigated lands in the Andijan region of Uzbekistan after people started to irrigate surrounding areas up slope. Cotton yields decreased from 30–35 centners/ha in the 1960s and 1970s to 20–22 centner/ha in the 1980s, and it is possible to give many such examples.

Unfortunately, specialists engaged in developing new irrigation systems also did not consistently implement scientifically recommended practices. Research in Turkmenistan, Uzbekistan and other regions demonstrated that increasing salinization was caused both by the construction of an insufficient network of water collection and drainage facilities and by the uniform application of technologies that were unsuited to the diversity of landscape and ecological conditions. Deterioration of these facilities and uncontrolled water use over a protracted period of time has also contributed to the problem.

Human population growth has exacerbated the situation. The population in the Central Asia in 1913 was 7.27 million people, in 1980 26.8 million, in 1990 33 million, and now is over 45 million people, i.e. the population has grown by 5–6 times in the past century. This growth has taken place in arid and semi-arid areas prone to desertification. For example the population of Uzbekistan grew by 3.5 times between 1913 and 1979. In Turkmenistan, in

1913 the population was 1.04 million people and in 1995 4.48 million people – an increase of more than four-fold in less than a century.

Feeding this expanding population was possible only on the basis of further expansion of the area of irrigated land or improvements in the system of irrigated agriculture. Since technical improvements gave results only after a period of time, primary attention was concentrated on the development of virgin and fallow lands, through the construction of dams, water reservoirs for regulation of water flow, and canal construction to divert the flows from the Amu Darya and Syr Darya into the basins of other rivers with low water or to arid districts. Reconstruction and improvement of the irrigation network took place, but it was insufficient to prevent large losses of water from canals due to leakage and infiltration.

Thus, the potential capacities of land and water resources for many years were used extremely irrationally and the condition of the land deteriorated due to technical mistakes, dictatorial approaches to choosing the strategic direction for water development, and agricultural mismanagement. The result has been the aggravation of the ecological situation and declining living standards.

2. Socio-economic consequences of desertification processes in Central Asia

Desertification in Central Asia is becoming more serious and more extensive.

The Aral Sea catastrophe has had a negative impact on the living conditions and the quality of life of 45 million inhabitants of the Aral Sea basin. High levels of unemployment, declining incomes, out-migration, reduced life expectancy, and increased child mortality are some of the indicators of this continuing ecological and social catastrophe.

Many publications are devoted to the problem of the Aral Sea; a number of regional and international meetings and seminars have addressed the topic, and special organizations have been established to deal with the problem. At the present time it is appropriate the think of the entire zone as a region in ecological crisis, which involves not only desert districts and the area adjoining the Sea itself, but also oasis, plains and nearby mountains. It is possible to predict that with time this crisis will continue to grow and that its coverage will expand.

From the perspective of the region's natural ecology, the vast irrigated areas of Central Asia constitute a qualitatively new natural-technological system with unknown properties. Due to the revolution in agriculture that has occurred in the last forty years, the region has been turned into a large source of salt, dust and moisture. What should be recognized is that the ecological and socio-economic conditions in the Aral basin are deteriorating across many parameters.

For over 60% of irrigated land in Central Asia the level of underground water exceeds the critical ceiling of 2 meters. As a whole the Central Asian region is therefore characterized by very high levels of salinity for irrigated land. At present, out of 8 million hectares of irrigated land in the Aral basin, over 60% is salinated to some degree and around 70% of this land is naturally poorly drained and requires complicated technical improvements if it is to be used on a sustainable basis.

According to the opinion of the majority of specialists, the most acute problem of the region is secondary salinization of lands and, related to this, the diversion of mineralized agricultural drainage water. However, out of 8 million hectares of irrigated land only 4.5 million hectares are provided with a system for collecting and disposing of used irrigation water. Drainage discharges into the Amu Darya and Syr Darya have resulted in mineralization which reaches more than 2 g/liter of water during selected periods in the middle and downstream sections of these rivers. The discharge of drainage waters into natural lowlands has created 2,300 lakes in the Aral Sea basin covering a total area of 8 thousand km^2.

The most intensive development of irrigated land took place in the 1970s and 1980s, when around 1.4 million hectares of new land were developed, which is why gross agricultural output increased. However, as the result of the extensive nature of these developments, the salinization of agricultural land, and population growth, agricultural production per capita decreased from 1980 to 1993 by 24%. This circumstance aggravated food supply problems, increased dependence on imported foods, and promoted unbalanced diets dependent upon cereals (Alibekova, 2000). Total loss of agricultural output from irrigated land in Uzbekistan has been 30%, in Turkmenistan 40%, in Tajikistan 18%, in Kazakhstan 30%, in Kyrgyzstan 20% (Glazovsky and Orlovsky, 1996). These figures allowed the American scientist H.E. Dregne to estimate in 1986 that in Central Asia 70% of land already could not be improved.

One of the main problems of irrigated agriculture of the Central Asia is the environmental impact of discharged irrigation drainage water. Irrigated agriculture takes around 90% of the flow of the Amu Darya and Syr Darya. At this rate of extraction, the volume of drainage waters is over 35 km^3. On the basis of the analysis of aerospace data it has been established that during last 30 years in Central Asia agricultural drainage water has flooded around 800 thousand hectares of land, effecting 930 thousand hectares where fodder crops were replaced by low value grass. For example, due to declining water availability, the area of haylands between the Talas and Assa rivers decreased by 75%, while the productivity of pastures was reduced to 1–1,8 centner/ha. As a result the cost of producing livestock products – meat, wool and astrakhan fur – sharply increased. Productivity of saigas decreased from 18 kg per animal in 1961 to 15 kg in 1984.

One of the most dangerous phenomena which lead to various socio-economic consequences in recent decades is a significant increase in atmospheric dust and the development of powerful sand and salt storms.

Drying of the Aral Sea has lead to the creation of large territories of open sea bottom, which are enriched with salts, fertilizers, pesticides, representing a very strong and virulent mix which is dangerous for people and for the environment. Small particles are very soluble and prone to wind erosion. It was calculated that annually 70 million tons of salts are taken off the Aral Sea basin and deposited over an area of one and a half to two million km^2, damaging farms significantly. Agricultural regions, such as Tashauzski and Karakalpakstan, located close to the former shores of the Aral Sea, have in particular experienced the impact of salts from this source.

Intensiveness of dust deposition depends first of all on the level of remoteness from sources of withdrawal, but numerous observations of dust on the surfaces of mountain glaciers are evidence of its intensiveness. The pollution of snow cover in the mountains of the Central Asia covers all high altitude zones, including the highest peaks. Increasing pollution on the surface of mountain glaciers acts as a melting catalyst. As a result, from 1959 to 1980 glaciers in Central Asia decreased in size by 19%.

Thus the problem of desertification in Central Asia is more serious than previously thought. 60% the territory of Uzbekistan and Kazakhstan and 66% of the territory Turkmenistan are prone to anthropogenic desertification.

Now we will briefly examine the processes of desertification, the ecological situation and their socio-economic consequences within individual republics in Central Asia.

3. National trends

Kazakhstan is a country especially prone to desertification. For Kazakhstan the coincidence of unfavorable natural and anthropogenic impacts has accelerated desertification. At the present time 63 million ha. of degraded pastures have been reported, which makes up over 30% of total pastures, as well as significant areas of pollution caused by nuclear test sites in the country. In the 20 million ha designated for the military, over 500 nuclear tests were carried out. According to expert opinion, the impact of military-space complex and effected not less than 1 million people (Yakovlev and Yakovleva, 1996). Arable land has lost 20–30% of its humus, 12 million ha. are subject to wind erosion and 50% of irrigated land is prone to secondary salinization.

In Kyrgyzstan processes of soil erosion cover an area of 5.5 thousand km^2 and the salinization of irrigated land effects one thousand km^2. As a result of pasture degradation, grassland productivity has decreased by 30–40%. In the last half century the forested area has declined by half and today the total area of forests is less than 4% of the territory of the country.

In Tajikistan secondary salinization and erosion have been observed in many agricultural areas and the productivity of pastures has decreased by 40–60% in the last 25 years. Areas of eroded land in the country comprise 98% of total area of agricultural land. Land slides are increasing in mountain districts due to economic activities such as tree cutting, uncontrolled irrigation, and the leakage of water from dams.

One of the most important ecological problems of ***Turkmenistan*** is salinization of soil as a result of irrigation and dam construction, which has caused the water logging and created alkali soils. Almost 90% of irrigated land of Turkmenistan presently is in unsatisfactory condition due to salinization. Total average annual accumulation of salts on the plains of Turkmenistan, for instance in the area of the Karakum channel, totals 9–10 million tons per year.

Ninety-nine percent of land is irrigated predominantly through surface irrigation through furrows. Total losses of water in irrigation systems comprise around 12 km^3, i.e. a half of total volume of water used in the country. The greater part of this water is lost through filtration, as the result of which large areas of land along channels become useless. Losses from the Karakum Canal are particularly high and are visible from the air because the numerous lowlands filled with water can be easily identified due to dense growths of reeds. For many kilometers back from the canals valuable fodder crops have been replaced by poor species. Thus, agriculture in Turkmenistan is currently facing and in will face in the future desertification problems in irrigated zone.

For the republic ***Uzbekistan,*** over 80% of the territory of which is occupied with deserts and semi-deserts, the issue of combating desertification is very relevant. Due to human activity 60% of the territory of Uzbekistan has been exposed to desertification of various stages, which damages the national economy.

The bulk of all crops (95%) are produced on irrigated land, which occupies 15% of all cultivated land of the country. The area of irrigated land in the republic is 4.3 million ha. and Uzbekistan spends around 90% of its available water resources on irrigation. According to data collected by the Uzdaverlioha Institute, salinated lands in the republic in 2003 comprised 66% of agricultural land (24,463 thousand ha.) and very-salinated land areas constituted around 35 percent. Cotton productivity decreases from 32–42 centners/ha. to 18–25 centner/ha. with increasing salinization. Soil salinization also has a negative impact also on quality of raw cotton. Annual economic losses due to salinization are large and are estimated at around 1,000 million USD (Mahmudov et al., 2002).

It is necessary to emphasize special problems of the Syr Darya and Dzhizak regions of the Republic Uzbekistan, where ecological conditions have worsened dramatically during the last decade. Areas of low and medium levels of salinization expanded from 30–50% in 1991 to 75–80% in 1999. This deterioration is related to the transformation in 1991 by Kyrgyzstan of the hydrological regime of the Toktogul water reservoir from production of water for irrigation to energy production, which has shifted the timing of the Syr Darya floods from spring to winter.

Thus, the main cause of salinization of soils and decreased productivity of the most important crops in the Syr Darya and Dzhizak regions of Uzbekistan has been increased soil water logging as a result of the flooding of the Aidarkul lakes and Syr Darya River in the autumn and winter period. The total economic damage for Uzbekistan has been over one billion USD (Mahmudov et al., 2002).

Ignoring local landscape in the design of irrigation schemes has also contributed to elevated groundwater levels and salinization in areas characterized by poor natural drainage. For example, the level of ground water is 1.5 meters for 70% of irrigated land in downstream Amu Darya areas, and for 50% of irrigated land along downstream sections of the Zarafshan River.

Inappropriate crop rotations also contributed to the problem. The area of land under perennial legumes in crops rotations was reduced in favor of cotton monoculture. This required excessive irrigation, and the heavy use of fertilizers and led to the gradual reduction of the humus content of the soil to 40–50% over the course of 40 years. Starting in the 1980s, these processes have produced declining crop yields and an estimated total yield loss of 30% in Uzbekistan.

Pastures in Uzbekistan occupy 85% of agricultural land or 23 million ha. and are the main source of support for over 10% of population. However, the economic and productive potential of these pastures is not realized. As a whole across the country over the last fifteen years the productivity of pastures has decreased on average by 23%, and average output of dry edible forage for grazing has been reduced from 2.4 to 1.8 centner/ha.

Pollution of natural environment of Uzbekistan due to excessive use of chemicals in the 1970s and 1980s has produced what could be called chemical desertification. In this respect Uzbekistan led the former Soviet Union. Around 80 different types of preparations were used in cotton production, at levels of up to 50 kg and over per hectare, which were fifteen to twenty times the average application rates in the USSR. The huge costs for the use of agricultural chemicals and expanding the area under irrigation were not economically justified, and also did irreparable damage to the environment and to the health of the population. Before the collapse of the USSR, Uzbekistan had the highest morbidity in the Union; child mortality exceeded 32%, and was even higher in the worst effected areas of the Aral basin. From 1990 to

1993 the use of pesticides was reduced by half and fertilizer by a third, but the benefits were slight because of the already high levels of environmental pollution. In recent years the deteriorating ecological situation in Uzbekistan has lead to a dramatic rise in morbidity among children and women, due largely to insufficient supplies of good potable water.

Ecologically driven population migration is a social and economic consequence of worsening of environmental situation in Uzbekistan. For example, the drying of the Aral Sea forced the out migration of several thousands of families from the Muinak District of Karakalpakstan, reducing the district's population from the mid-1970s to the mid-1980s from 45,000 to 22,000 people (Glazovsky and Orlovsky, 1996). This pattern has continued up to the present. Shortage of water during 2000–2001 due to drought sharply decreased the living standards of the population, especially in the northern districts of Karakalpakstan, and many people left the area.

Migration caused by ecological problems has also been observed in the downstream along the Zarafshan River (Bukhara region), where desertification was acute and annually tens of thousands of people are leaving agricultural districts and relocating in towns. A needs assessment of the rural population in this area showed that most income (80–85%) was spent on buying food.

Environmental deterioration has also effected infrastructure. Soil salinity has reduced the period of service of poles for high voltage electric power lines, leading to frequent interruptions in electrical service and high repair costs. The mineralization of groundwater has also caused the deterioration of the foundations and walls of many buildings constructed only 30–50 years ago. It has also threatened the ancient archeological monuments of the lower Amu Darya. According to data collected by S.Kamalov (1998), until recently in the Ellikkalansky District there 50 ancient monuments, while now only ten remain. The observatory of Koikyrylgan-Kala, attributed to the IV century B.C., was destroyed completely in Turtkulsky District (Kamalov, 1998).

4. Conclusion

The risk from desertification is very serious for countries in Central Asia, as a result of many years of economic development within the USSR and without consideration for the long-term ecological consequences of these developments. Poor environmental conditions in the region have undermined living standards.

At a time when many parts of the world are subject to social and interethnic conflicts, environmental threats are frequently seen as a secondary problem. Nonetheless, desertification in Central Asia covers vast areas and is growing despite investment in environmental improvement activities. In our opinion, numerous environmental protection programs and projects have been unsuccessful because they were based on faulty premises. Up to the present, successes in combating desertification in Central Asia are uneven and unsatisfactory, and in a number of districts the situation even worsened.

This is why it is very important to review former assessments, approaches and actions, despite their limitations. Practice shows that private activities alone do not produce an appropriate result in fighting against desertification. Only a comprehensive approach to the economic development of an entire country and region based on scientific geographical, ecological and socio-economic research will open opportunities for the management of a complex natural process like desertification, which is caused by number of natural and anthropogenic factors.

Desertification takes place within naturally delimited geo-systems; it must also be studied and prevented within the framework of these natural geo-systems. In order to achieve reliable results in combating desertification it is necessary to raise the level of knowledge and ecological thinking. As Academician V.I. Vernadsky demonstrated, the status of nature in the modern world is increasingly determined by human knowledge.

For an understanding of the dynamics of desertification, as well as for the development of ecological and development policy, information and systematic observations must be coordinated on global, regional, national and local levels, and based on material provided by scientific research. Hence, the role of research increases, while science itself becomes an extremely important component in the search for ways to provide for sustainable development in conditions of vulnerable ecosystems.

States of Central Asia, share deep cultural, historical and economic commonalities as well as a single ecological space, are connected by common water arteries, and – being located in the arid zone – are in practice vulnerable to mistakes of an ecological nature. A significant part of our natural resources has been exhausted, while the ecological situation becomes more and more unfavorable, and it is therefore necessary to undertake urgent measures.

References

Alibekova, S. L., 2000, Living standards of the population of Uzbekistan in conditions of desertification, *Issues of desert development* **3**.

Glazovsky, N. F., and Orlovsky, N. S., 1996, Issues of desertification and draught in the CIS. Ways for their resolution, *Issues of desert development*, **2**.

Kamalov, S., 1998, Process of migration in Karakalpakstan, *Issues of desert development* **3,4**.

Mahmudov, E. J., Bazarov, D. R., Islamova, N. K., and Kuchkorova, D.H., 2002, Experience and issues of the use of water resources in Uzbekistan, *Bulletin of the Mountains of Central Asia*, **2**.

Yakovlev, V. A., and Yakovleva, G. G., 1996, Technogenic aspects of the natural environment of Kazakhstan, *Issues of desert development*, **1**.

PART II

GRAZING SYSTEMS AND DESERTIFICATION

CHAPTER 4

FORAGE DISTRIBUTIONS, RANGE CONDITION, AND THE IMPORTANCE OF PASTORAL MOVEMENT IN CENTRAL ASIA - A REMOTE SENSING STUDY

IMPORTANCE OF PASTORAL MOVEMENT IN CENTRAL ASIA

MICHAEL COUGHENOUR*[1], ROY BEHNKE[2], JOHN LOMAS[3] AND KEVIN PRICE[3]

[1] *Natural Resource Ecology Laboratory, Colorado State University, Fort Collins, Colorado, 80523, U.S.A.*
[2] *Macaulay Institute, Aberdeen, AB15 8QH, Scotland, U.K.*
[3] *Kansas Applied Remote Sensing Program, University of Kansas, Lawrence, Kansas, 66045 U.S.A.*

Abstract: Pastoral ecosystems in Central Asia have been undergoing many changes, with important implications for their sustainability. Pastoralists traditionally moved over large distances, either following regular migrations between seasonal pastures or opportunistically following forage; however, the ranges of pastoral movements have been reduced. Remote sensing methods were used to assess 1) whether these changes in pastoral land use have promoted or reversed affected rangeland degradation, and 2) how spatial and temporal variations in forage biomass might affect livestock movements and productivity. The remote sensing approach involved measurements of an index of plant productivity, the NDVI, over broad geographic areas. The relationships of plant production to precipitation over time and space were examined in order to differentiate the potential effects of precipitation from the potential effects of livestock grazing on rangeland condition. This was accomplished by assessing "rain use efficiency" or "RUE", which is an index of the amount of plant production per unit of precipitation. Four study sites or regions were assessed; two in Turkmenistan and two in Kazakstan. Each of the four case studies revealed a different situation with respect to

* To whom correspondence should be addressed. Michael Coughenour, Natural Resource Ecology Laboratory, Colorado State University, Fort Collins, Colorado, 80523, U.S.A. e-mail; mikec@nrel.colostate.edu

R. Behnke (ed.), The Socio-Economic Causes and Consequences of Desertification in Central Asia, 45–80.

the distribution of resources, the importance of movement, the degree to which movements have been altered, and the consequences for rangeland condition. The analyses revealed spatio-temporal patterns of precipitation, forage, water, and topography which necessitate movement and adaptability. Although traditional regional scale migrations have been lost, the analyses suggested that smaller scale movements, coupled with appropriate stocking rates, can partially avert the negative consequences of complete sedentarization for pastoral production and rangeland condition. To maintain or improve pastoral productivity and sustainability in Central Asia, it will be essential to integrate the spatial distributions of forage, water, climate, and the effects of alternative livestock movements on the condition of the livestock as well as the condition of the rangelands.

1. Introduction

Pastoral ecosystems in Central Asia have been undergoing many changes, with important implications for their sustainability. In Kazakstan and Turkmenistan, pastoralists traditionally moved over large distances, either following regular migrations between seasonal pastures in a transhumant system, or opportunistically following forage in a more nomadic fashion. The ranges of pastoral movements have been reduced in both countries. During the Soviet era, beginning in the 1930's, pastoralists were sedentarized into a collective system of state-run farms. Livestock production systems were intensified, including the provision of state subsidized fodder production and distribution systems. The intensified, sedentarized systems likely experienced higher stocking rates, and rangelands likely had fewer rest periods for regrowth, possibly leading to rangeland degradation in some areas.

When Kazakstan became independent, during the break up of the former Soviet Union, the system shifted from a collectivist to a privatized, open market system (Kerven 2003, Behnke 2003). State subsidies for fodder, transportation, and maintenance of wells ended. As a result, livestock populations declined markedly between 1993 and 1998. Livestock had to be sold, and rangelands destocked, because the pastoral systems could not be sustained economically. Furthermore, it was not possible to revert to the traditional transhumant or nomadic systems had long ago been disrupted as pastoralists had been sedentarized and traditional grazing rights had been lost. There is evidence that in response to the decline in livestock numbers in Kazakstan, there has been a trend of increasing rangeland condition compared to the degraded state of many areas during the intensified livestock production systems of the Soviet era (Alimaev 2003).

In contrast to the rapid shifts in Kazakstan after independence, the transitions in Turkmenistan have been more gradual. Livestock production systems and grazing lands were transferred to leasing associations and private individuals (Kerven 2003, Lunch 2003). The leasing system has provided pastoralists with economic opportunities to build up and maintain their herds. Grazing land and herds are held by farmers associations. The associations lease animals to individual herders, who meanwhile try to obtain their own livestock through their incomes. The leasing system is a step-wise approach to privatization and pastoralists are increasingly faced with economic pressures as government support for the associations is withdrawn.

As in Kazakstan, however, livestock mobility has been reduced. During the Soviet era, traditional long-range movements were abandoned, and nomads were encouraged to settle around service centers in the desert (Khanchaev et al. 2003). Collective state-run farms were formed, and they were subsized with supplementary winter feed, veterinary services, and transport. Movements of sheep and camels to northern pastures were also subsidized by the state. In particular, sweet water from the Kara Kum Canal was trucked into the northern desert, allowing pastoralists to use that area in the winter. Wells in that area are also falling into disrepair. With the loss of state support following the Soviet breakup, many pastoralists are no longer able to make the migratory moves to northern pastures in the winter. An increasing threat is the potential loss of the remaining government support for the farmers associations due to the instability of the Turkmenistan economy (Kerven 2003).

Due to the state subsidized development, more people and livestock were concentrated near settlements. This resulted in degradation near the villages (Khanchaev et al. 2003). Since the reforms in the state farm system in the 1990's, increasing privatization, reduced state subsidies have placed financial stress on pastoralists, making it more difficult for them to move longer distances, thereby resulting in further concentrations of livestock near settlements. Meanwhile, the reduction in water supplies in the northern deserts has resulting in reduced grazing pressure in those areas.

The objectives of this study were to use remote sensing methods to assess 1) whether these changes in pastoral land use have promoted or reversed affected rangeland degradation, and 2) how spatial and temporal variations in forage biomass might affect livestock movements and productivity. The remote sensing approach involves measurements of an index of plant productivity over broad geographic areas. The study will also examine how plant production is related to precipitation over time and across the landscape, in order to differentiate the potential effects of precipitation from the potential effects of livestock grazing on rangeland condition. This will be accomplished by assessing "rain use efficiency" or "RUE", which is an index of

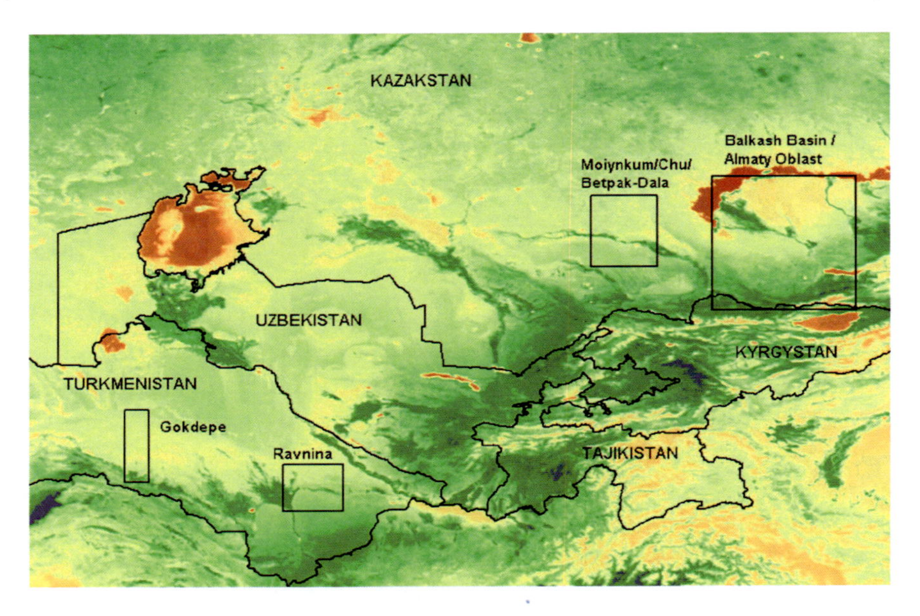

Figure 1. Locations of the four Central Asia study areas in Turkmenistan and Kazakstan and country boundaries overlain on a map of annual average NDVI for 1989–2003. Green means more green biomass

the amount of plant production per unit of precipitation. If a rangeland has been degraded, RUE will be reduced.

Four study sites were assessed on the DARCA (Desertification and Regeneration: Modelling the Impact of Market Reforms on Central Asian Rangeland) project, two in Turkmenistan and two in Kazakstan (Figure 1). These study sites will be examined here on a case by case basis, following a description of the overall methodology.

2. Methods

The normalized difference vegetation index (NDVI) was used to assess green biomass. The NDVI is computed from two spectral bands as

$$\text{NDVI} = (\text{Near Infrared - Visible Red})/(\text{Near Infrared + Visible Red})$$

This vegetation index has in use since the early 1980's globally, across a wide variety of vegetation types and levels of green biomass, from deserts and desert margins in the Sahel, to tropical rainforests (Tucker and Sellers 1986, Justice et al. 1998, Prince et al. 1998). Green vegetation has a high reflectance in the near infrared band and a low reflectance in the visible band of the spectrum. Thus, the greater the difference, the more green biomass. The difference is normalized to the sum of the reflectances to scale for differences in total reflectance caused by variations in illumination intensity.

NDVI has usually been computed from the NOAA Advanced High Resolution Spectroradiometer (AVHRR) instrument on board a series of NOAA polar orbiting satellites since 1982. The satellites orbit the globe many times every day while the earth revolves below it, thus providing global coverage every day. Daily data are usually composited on a 10 day or 15 day basis to obtain a single scene with minimal cloud cover. The resolution of the data is approximately 1 km × 1 km. These data are referred to a LAC or local area coverage data. NOAA also aggregates the data to approximately 4 km × 4 km to provide a global area coverage or (GAC) data set, which may also be obtained from NOAA.

To achieve the aims of this study, it was necessary to use the higher resolution 1 km LAC data. The Kansas Applied Remote Sensing Program (KARS) obtained LAC data from NOAA. For 1982–1984 the only AVHRR data that were available were the GAC data. For 1985, NOAA had a limited set of LAC scenes. LAC data were available from NOAA for 1985–1988, but they had to be purchased at a prohibitive cost. Data from 1989-present were available on line from the NOAA data access center at no cost. Thus, the current study covers the period 1989–2003.

The data that were downloaded included that portion of Central Asia which included all four of the DARCA study areas (Figure 1). After downloading the data from the NOAA site, KARS used the techniques used by EROS Data Center to correct for water vapor, ozone, and surface geometry, and to then compute the daily NDVI scenes. The daily scenes were composited biweekly. The biweekly composites were georectified to a common map base using a manual method of selecting ground control points and building a transformation model to "rubber sheet" each image to a common geographic coordinate system (latitude and longitude). KARS then used a set of algorithms to compute the vegetation phenology metrics "date of onset of greenness" and "data of maximum NDVI". Spatial coverage in the NOAA AVHRR data were not 100% In particular, data for 1989, 1990, 2000, 2001 were missing for important portions of Kazakstan (Coughenour 2005). In the year 2000, the NOAA 14 satellite orbit was degrading and the NOAA 16 satellite did not come on line until March 2001. The biweekly NDVI, and the two phenology metrics based upon all available LAC data from 1989–2003 comprised the final data set that was delivered to Colorado State University.

The scenes were then converted to the IDRISI geographic information system. The scenes were further georectified to a higher level of accuracy against hydrological ground control points in the hydrology layers for Turkmenistan and Kazakstan in the Digital Chart of the World (ESRI). This process involved using mathematical transformations available in the IDRISI "resample" procedure.

Country boundaries, roads, and hydrology coverages were obtained from the Digital Chart of the World (ESRI) and converted to IDRISI. The digital elevation model used in the precipitation interpolations was based on the 5-minute global DEM ETOPO5, created by NOAA and available at the National Geophysical Data Center.

Annual precipitation data were obtained for weather stations in each of the four study areas from local sources. There was nearly complete coverage from these stations for the entire period 1989–2003. Additional data for weather stations in Kazakstan and Turkmenistan were obtained from the Global Historical Climatology Network (GHCN).

The precipitation data were used in a spatial interpolation program (PPTMAP, Coughenour unpublished) to generate annual precipitation maps. The interpolation program is based upon algorithms first developed for use in the SAVANNA ecosystem model (Coughenour 1992). The algorithm uses inverse distance weighting, correcting for the elevation difference between the weather station and the point in question with a linear regression of precipitation on elevation in the local area.

Rain use efficiency (RUE) maps (Prince et al. 1998) were generated by dividing the average NDVI map for February-September by the annual precipitation map. The average NDVI is viewed as being mathematically equivalent to the time integral of NDVI, which is theoretically related to total intercepted radiation by green leaves, and thus net primary production.

3. Case studies

3.1. THE GOKDEPE STUDY AREA, WEST CENTRAL TURKMENISTAN

The Gokdepe area of west central Turkmenistan consists of a north-south transect approximately 200 km long from a zone in the south with settlements near the Karakum canal north into the desert (Figure 1). The DARCA project surveyed 20 villages along the transect, holding 23,000 sheep and 2,500 camels in total. Most of the sheep are based in the south and move north-south every year. Winter wells, which are saline, are located mostly in the north. Summer wells, apart from those in settlements, are fresh, and are located mostly in the south. Sheep in the north also move east to west in response to water salinity. Permanently settled communities in the north are located near fresh water sources where animals summer while their water requirements are high. Forage is depleted in the settled areas during summer, at which time they move outwards to utilize more saline wells in the winter, where grazing pressure is less intense. Thus, one might expect that the areas around settlements could be heavily grazed and thus degradated (Behnke et. al, this volume).

The major questions at this site are:

1. How does forage biomass vary seasonally along the north-south gradient of livestock movement?
2. Is there evidence of decreased range condition in areas near settlements or wells?
3. Have forage productivities and rangeland conditions improved or degraded over time since 1989?

The study area is located in one of the lowest elevation regions of Central Asia. Elevations range from 60–150 m for the most part (Figure 2). Elevations begin to increase in the far south to approximately 600 m. Most of the area is covered with deserts, plains and melkopsopochnik (a type of denuded relief comprised of hills and depressions). In the southwest there are peidmont and mountain deserts, savannoids, and a small amount of phryganoids (open woodlands and shrublands). In the southwest, irrigated agricultural fields line the Karakum canal where it runs through the study area. Mean annual rainfall for 1989–2003 for this study area ranged from approximately 102–114 mm in the northeast to approximately 165 mm in the southwest. Most of the region received less than 150 mm per year on average. Mean rainfall in the most prevalent vegetation zone (deserts and plains) was125 mm with a CV of 0.25

Mean annual NDVI distribution largely followed the precipitation distribution (Figure 3A), with higher NDVI in the southwest. However, NDVI was

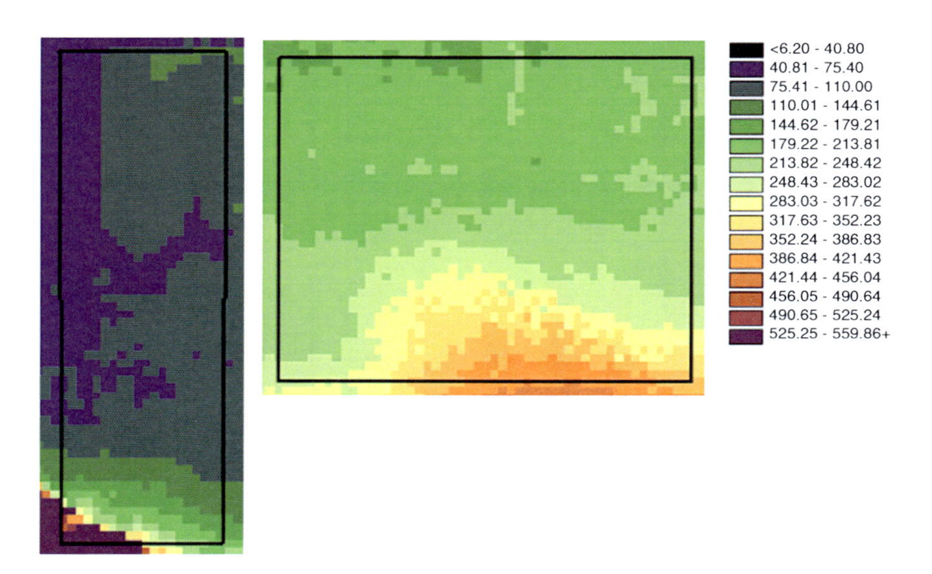

Figure 2. Elevations (meters) of the DARCA Turkmenistan sites. Gokdepe (left), Ravnina (right)

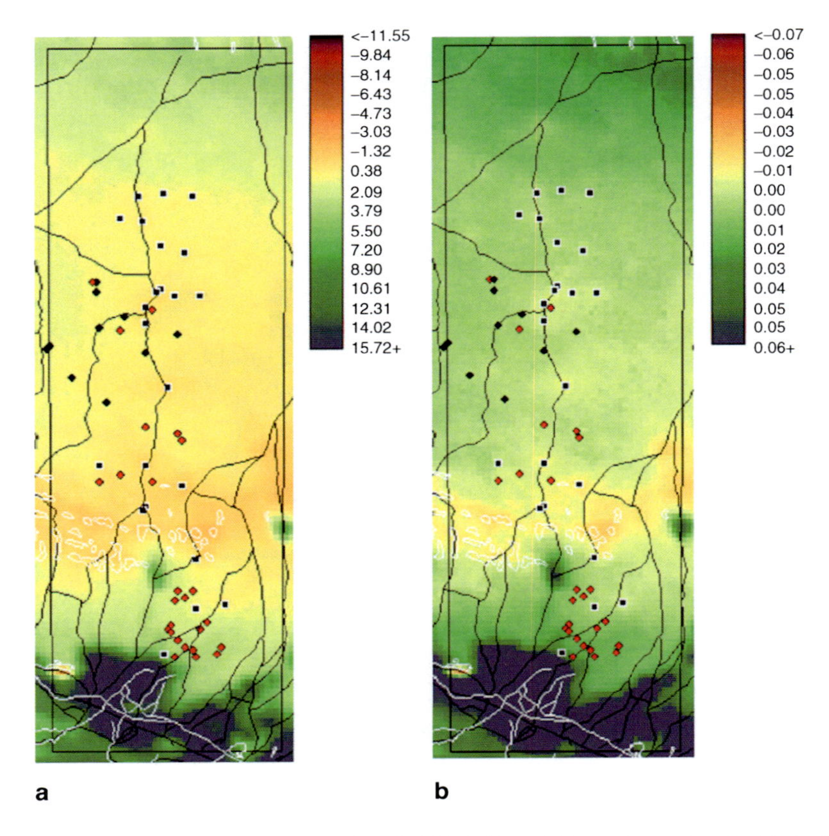

Figure 3. Gokdepe site. A) Mean annual NDVI*100 1989–2003 with roads (black), hydrology (white), villages (black squares), winter water sources (black), and summer water sources (red). B) Mean RUE 1991–2003. The white irregular polygons are Solochaks (hypersaline) depressions

not lower in the dry area of the northeast. The effect of irrigation along and near the Karakum canal is evident in the southwest. The NDVI is elevated along the canal, then it decreases further to the southwest while precipitation continues to increase in that direction. Rain use efficiency (RUE) was highest in areas with the highest NDVI (Figure 3B). RUE was markedly higher in the vicinity of the Karakum canal, due to irrigation and ground water subsidies (Figure 3B). There was a notable band of low RUE running east-west across the central part of the study area (Figure 3B). This is an area of hypersaline desert and solochak depressions. There were some areas near villages that appeared to have lower RUE. This was less evident near the villages in the far north and far south.

NDVI and RUE values were examined in relationship to villages along a north-south gradient (Figure 4). Village grazing areas were defined as the area within a 10 km radius of the village. The village areas were ascribed to

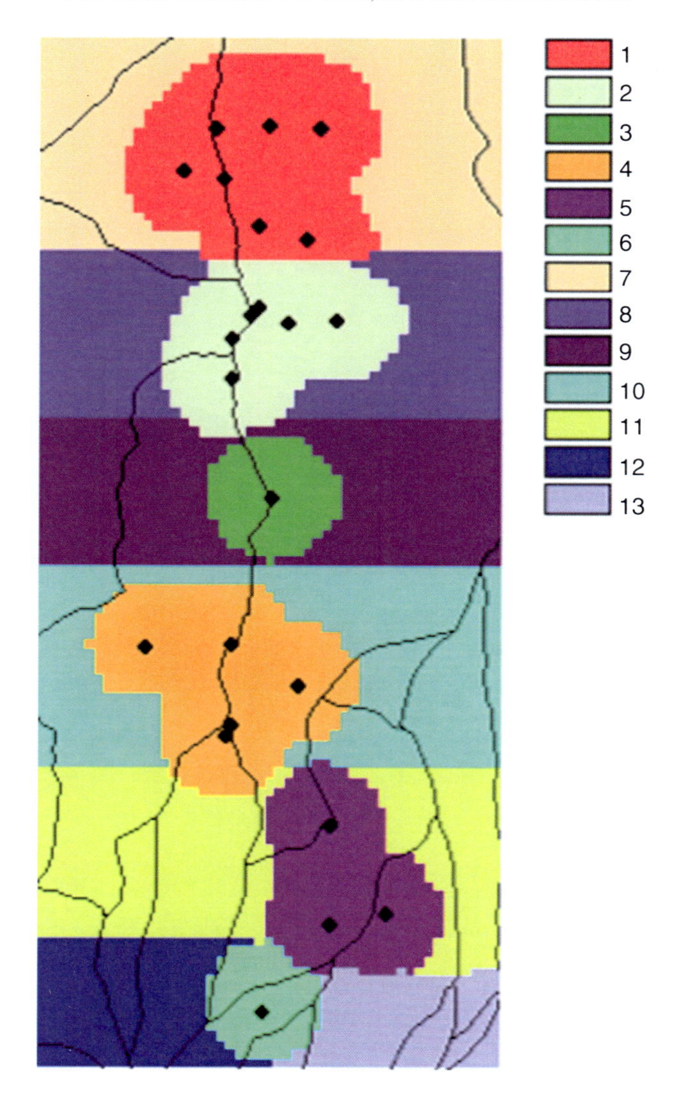

Figure 4. North-south village grazing areas (1–6) and zones outside of the village grazing areas (7–13) in Gokdepe, Turkmenistan

6 north-south zones, numbered from north to south. The adjacent areas outside the village grazing areas were also numbered from north to south. The southernmost area outside the village grazing area was split in two, because of obvious differences in NDVI on the east vs. west sides.

The mean NDVIs outside village grazing areas 1–4 were notably higher than inside the village grazing areas (Table 1). NDVI was higher outside areas 5–6, but the relative difference was less marked. There were significant trends for increasing NDVI ($p \leq 0.1$) inside village grazing areas 1 and 2, and there were weak trend ($p < 0.16$) in areas 3 and 4. Outside the village areas, there were increasing trends outside areas 2 and 3 ($p < 0.1$), and there was a weak positive trend ($p = 0.11$) outside area 1.

RUE was higher outside all of the village grazing areas except that of village 5 (Table 2). It was 0.002 units higher outside of village areas 1 and 3, 0.003 units higher outside village area 2, 0.001 units higher outside village area 4, and 0.011 units higher outside village area 6. Unlike NDVI, there were no significant trends in RUE in the village areas or outside the village areas. There was a weak trend ($p = 0.014$) inside village area 1, and outside village areas 1,2, and 4 ($p \leq 0.016$). Thus, trends in NDVI were primarily caused by trends in rainfall (Table 2).

There were significant ($p < 0.1$) positive trends in NDVI in all distance classes from winter wells (Table 3). Trends were positive but not as significant for villages and non-winter wells. Trends were less significant for RUE (Table 4). Thus, some of the trends in NDVI were likely due to trends in RUE, but most of the trends in NDVI were due to trends in rainfall.

Seasonally, NDVI increased from February through April in this region. The peak was in April except at village grazing area 6 and the area outside it

TABLE 1. Trend analysis and means of NDVI inside and outside of north-south village

	Inside Village Grazing Area			Outside Village Grazing Area		
Village Grazing Area	r2	P	Mean	r2	P	Mean
1	0.45	0.09	0.26	0.43	0.11	0.71
2	0.45	0.10	−0.12	0.46	0.09	0.12
3	0.42	0.12	−0.27	0.45	0.09	−0.60
4	0.38	0.16	−0.69	0.38	0.16	0.82
5	0.34	0.22	0.79	0.38	0.21	0.82
6	0.21	0.44	8.75	0.25	0.38	9.61

TABLE 2. Trend analysis and means of RUE inside and outside of north-south village

	Inside Village Grazing Area			Outside Village Grazing Area		
Village Grazing Area	r2	P	Mean	r2	P	Mean
1	0.39	0.14	0.001	0.39	0.15	0.003
2	0.37	0.18	−0.003	0.39	0.15	0.000
3	0.34	0.21	−0.006	0.38	0.16	−0.004
4	0.32	0.25	−0.009	0.32	0.24	−0.010
5	0.31	0.24	0.002	0.32	0.24	0.001
6	0.23	0.41	0.061	0.36	0.19	0.072

TABLE 3. Trend analysis and means of annual average NDVI*100 in distance zones from wells and villages in Gokdepe

Location Type and Distance	r2	P	Mean
Village 0–3 km	0.37	0.17	0.59
Village 3–5 km	0.40	0.14	0.70
Village 5–10 km	0.36	0.13	1.02
Summer well 0–3 km	0.38	0.16	1.56
Summer well 3–5 km	0.37	0.15	1.25
Summer well 5–10 km	0.40	0.14	1.35
Winter well 0–3 km	0.44	0.10	0.42
Winter well 3–5 km	0.45	0.09	0.38
Winter well 5–10 km	0.45	0.09	0.33

TABLE 4. Trend analysis and means of annual average RUE (NDVI/mm precipitation)in distance zones from Gokdepe wells and villages

Location Type and Distance	r2	P	Mean
Village 0–3 km	0.330	0.225	0.003
Village 3–5 km	0.359	0.190	0.003
Village 5–10 km	0.382	0.160	0.006
Summer well 0–3 km	0.359	0.189	0.009
Summer well 3–5 km	0.362	0.185	0.007
Summer well 5–10 km	0.382	0.160	0.008
Winter well 0–3 km	0.401	0.138	0.002
Winter well 3–5 km	0.404	0.135	0.002
Winter well 5–10 km	0.398	0.141	0.001

to the west (area 12). There was, on average, a decline in NDVI from April to a minimum in July, followed by a slight increase between July and September. The onset of greenness occurred later in the south, by as much as 90 days. This was likely due to higher elevations and thus colder and longer winters. Over most of the lower elevations, however, there was very little difference in green-up date. The data of maximum NDVI was also later in the southerly locations, in early to mid-May. However over most of the lower elevations there was a patchy assortment of maximum NDVI dates in February and March. Dates of onset and maximum greenness by village area revealed little differences amongst village areas 1–5. Village area 6 was clearly delayed, however.

3.2. THE RAVNINA STUDY AREA, EAST CENTRAL TURKMENISTAN

The Ravnina study area is also located along the Karakum, canal, but is at slightly higher elevation than Gokdepe (Figure 1). Elevations ranged from approximately 180 m in the north to just under 400 m in the south (Figure 2). Vegetation of the northern two thirds of the area is mostly comprised of deserts, plains, and melkosopochnik. The Karakum canal runs east and west, with meadow vegetation over much of its length, flowing into a large area of oasis vegetation in the southeast. There is an area of tugais (riparian woodland). The higher elevation areas in the south are mainly covered with savannoid mountain and peidmont vegetation.

Rainfall was estimated to be highest in the southwest, generally decreasing to the northeast. The gradient spans approximately 125–190 mm/year. Higher elevations occur in the south, so precipitation is likely to be greater. Temporally, there was a cycle of 5–6 years, with peak rainfall years in 1982 and 1998, and minimum rainfall years in 1995 and 2000. This was similar to the pattern observed in Gokdepe. Mean annual precipitation in the village, winter, and summer grazing areas was 133, 159, and 162 mm, respectively, with CV's of 0.26, 0.25, and 0.25, respectively.

The Ravnina study area has experienced increased sedentarization and intensified grazing near villages and water sources. The Ravnina village area is located just north of the canal between three village wells (Figure 5). There, private flocks of sheep are kept. South of the canal is an extensive area utilized by grazing associations. This area is defined by the locations of

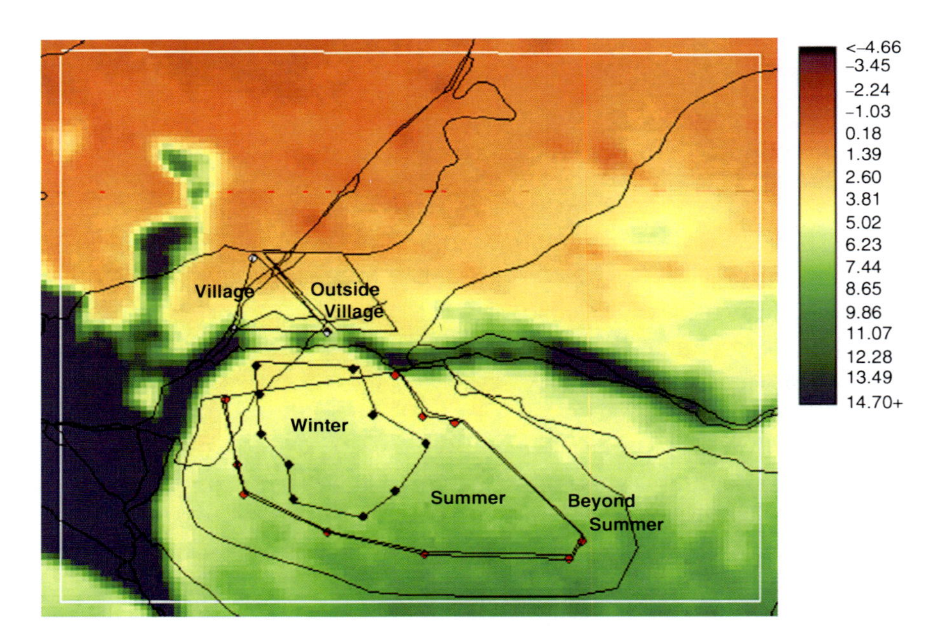

Figure 5. Mean annual NDVI*100 1989–2003 in Ravnina, Turkmenistan. Location of grazing areas are shown, along with village wells (white), winter wells (red) and summer wells (blue), and comparative sites outside the village and beyond the summer range

wells used previously in winter vs. summer. The winter grazing area is located just south of the canal, while the summer area extends outwards to the east, west, and especially to the southeast.

The primary questions at this study site are:

1. What is the seasonal forage availability at the three grazing areas (village, winter, summer)?
2. Are summer pastures better forage sources in summer?
3. Are private flocks kept north of the canal at a disadvantage with respect to forage?
4. Is there any evidence of degradation or recovery in any of these areas?

The area south of the Karakum canal had higher NDVI than the area north of it and NDVI increased gradually to the south (Figure 5). The meadows lining the canal had higher NDVI than other nearby areas due to the water subsidy. Similarly, the oasis and tugai areas had markedly higher NDVI's. The distribution of NDVI differed from the rainfall distribution in that the rainfall map did not show uniformly higher values across the southern portion of the study area. Nevertheless, precipitation was estimated to be greater in the area of summer wells than in the area of the village wells. RUE was higher in the south than the north and it was highest in the water subsidized areas (Figure 6).

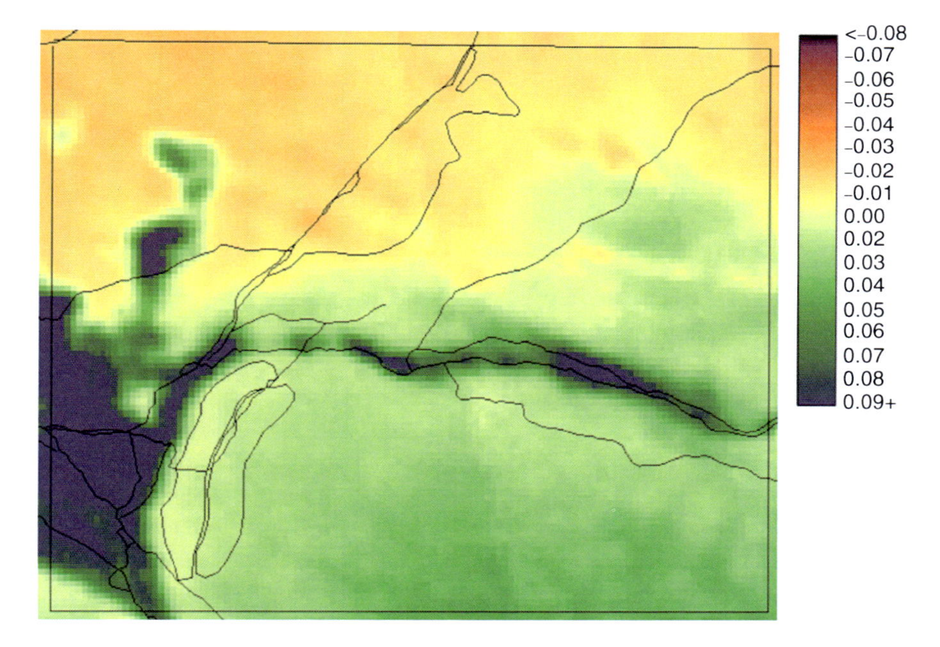

Figure 6. Mean RUE 1991–2003 in Ravnina. The two areas encircled in polygons are areas for which means are reported in the text

NDVI in the village grazing area was similar to that in the winter grazing area while NDVI in the summer grazing area was greater than in the village or winter grazing areas (Table 5). Differences in NDVI between these grazing areas were very small in comparison to differences among years. The summer grazing area is located in higher precipitation zones at higher elevations. RUE in the summer area was not proportionately higher than in the village or winter grazing areas (Table 6, thus, the higher NDVI in the summer grazing area could be attributed to greater precipitation.

NDVI in an adjacent area outside the village grazing area was slightly lower than in the village grazing area, the opposite of what might be expected if there was degradation in the village (Table 5). The difference could have been due to a lower RUE in the vegetation outside the village area. It is also possible that there was some irrigation in the village area supported by water from the canal. A comparison of the summer area and the area beyond the summer area showed no notable difference in NDVI or RUE, suggesting there were no negative effects of grazing in that area.

TABLE 5. Trend analysis and means of annual average NDVI*100 in various Ravnina locations

Location	r2	P	Mean
Village grazing area	0.32	0.26	5.92
Outside village grazing area	0.31	0.28	4.74
Winter grazing areas	0.35	0.22	5.94
Summer grazing areas	0.37	0.19	7.57
Beyond summer grazing areas	0.39	0.17	7.75
0–5 km from winter wells	0.38	0.17	6.36
5–10 km from winter wells	0.41	0.14	6.67
0–5 km from summer wells	0.40	0.16	7.01
5–10 km from summer wells	0.39	0.16	7.56

TABLE 6. Trend analysis and means of annual average RUE (NDVI/mm precipitation) in various Ravnina locations

Location Type	r2	P	Mean
Village grazing area	0.30	0.30	0.040
Outside village grazing area	0.31	0.31	0.035
Winter grazing areas	0.36	0.20	0.038
Summer grazing areas	0.33	0.25	0.047
Beyond summer grazing areas	0.37	0.20	0.046
0–5 km from winter wells	0.37	0.19	0.038
5–10 km from winter wells	0.41	0.15	0.041
0–5 km from summer wells	0.33	0.25	0.047
5–10 km from summer wells	0.38	0.18	0.046

A comparison of areas inside and outside the village grazing area on a monthly basis showed why the area outside the village area had lower NDVIs than inside the village area (Coughenour 2005). NDVI in both areas reached the same maximum value, but after April, the area outside the village area declined to a greater extent. A comparison of NDVI in the summer grazing area to areas outside the summer grazing area showed no differences.

Certain areas south of the canal had notably lower RUE (Figure 6). For example just south of the canal and just east of the oasis vegetation in the southwest, particularly near the roads, the RUE appeared to be lower than other areas to the south and east. RUE is notably lower in the small area east of the oasis that is encircled by roads. In the polygon east of the road RUE is 0.0065±0.0021 while in the polygon west of the road it is 0.011±0.0017. The lower RUE suggests that these areas may be in poorer condition.

There were no significant ($p < 0.1$) time trends in NDVI in any of the grazing areas (Table 5). However there were weak positive trends ($0.14 < p < 0.17$) in areas within 10 km of winter and summer wells. Similarly there were no trends in RUE in any of the grazing areas (Table 6), but there were weak trends for increasing RUE 0–10 km from winter wells and 5–10 km from summer wells. This would be a pattern that would be observed if range condition and basal plant cover were increasing.

An important conclusion to be drawn from the NDVI and RUE fluctuations is that a large portion of the fluctuations in NDVI was due to fluctuations in RUE, which was illustrated by the similarity of fluctuations of the two variables (Coughenour 2005). This is not to discount the effects of annual variation in precipitation, which also contributed to NDVI variations. However, since RUE tended to be higher in high precipitation years, there is evidence that higher precipitation contributes to increased NDVI in a complex, synergistic fashion by influencing both RUE, which can be considered a measure of rangeland condition or basal plant cover, and by influencing the production produced by that amount of basal cover.

Seasonally, the pattern was for NDVI to increase February-April, reaching a peak in April, followed by a gradual decline from April to July (Figure 7). Mean monthly NDVI values in February-May were considerably higher in the summer area than in the village area. NDVI in the winter area was intermediate. Interestingly, NDVI was not higher in the summer area in summer. Since NDVI only senses green biomass, it is likely that there was still more forage on the summer range in summer although most of it would have senesced. In the preceding spring period NDVI was higher on the summer range and, thus, biomass production would have been higher.

The date of onset of greenness increased gradually from north to south. In the village area, the onset date was approximately early February. On the winter grazing area it varied from early February to late February, while on the summer grazing area the onset occurred in early to mid March.

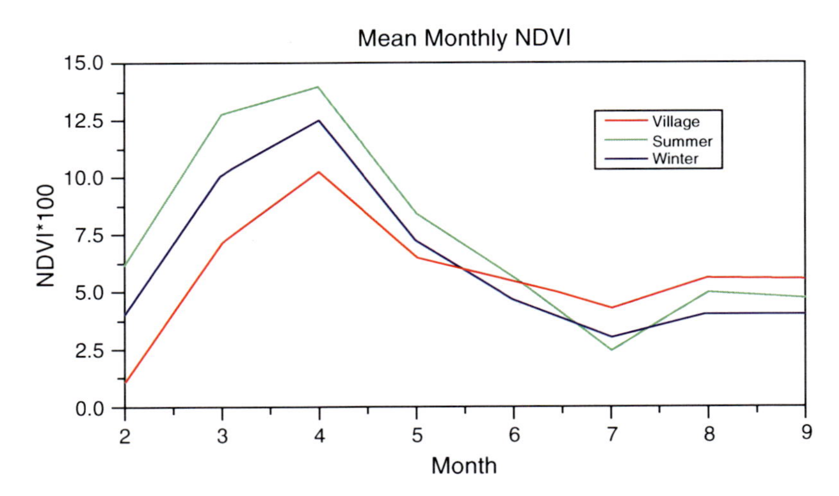

Figure 7. Average monthly NDVI for the three primary grazing areas in Ravnina

3.3. THE MOIYNKUM/CHU/BETPAK-DALA STUDY AREA, SOUTH-CENTRAL KAZAKSTAN

The traditional large-scale pastoral grazing system in the Moiynkum/Chu/Betpak-Dala area included seasonal north-south movements, with winter grazing in the southern Moiynkum desert, spring and autumn grazing in Betpak-Dala, and summer grazing further north (Robinson and Milner-Guilland 2003). These movements persisted until the mid 1990's in some places, and may be coming back in the next few years. The current system is much reduced in scale. However, there are still seasonal redistributions to take advantage of green forage when and where it becomes available. Grazing is heavily focused proximally to village centers, indicative of increased sedentarization and reduced mobility (Robinson and Milner-Guilland 2003).

Within the DARCA study area there are three villages located along the Chu River, which were headquarters for collective farms during the Soviet era. The villages were sited at spots which could support a large number of people and where flocks crossed the river on their annual migrations. The village of Sary Ozek is located in a broad floodplain which provides a wide variety of grazing resources throughout the year, while the village of Malye Kamkaly is located where the river flows through a single channel (Kerven et al. 2004). Furthermore, Malye Kamkaly is located on the north side of the river where there is only seasonal grazing. The village of Ulanbel is intermediate between these two extremes. Since independence in 1991, the human populations of Sary Ozek and Ulanbel fell but then stabilized and increased a bit. However, the population of Malye Kamkaly decreased steadily down to only a few families and the village appears to be dying. The few remaining families persist by practicing large scale seasonal migrations (Kerven et al. 2004).

Research by DARCA investigators has shown the following (Behnke and Temirbekov 2005). In Sary Ozek, a significant number of livestock are able to graze near the village all year long (50% of 2,800 sheep and goats), and the village is thriving. In Malye Kamkaly, most of the remaining livestock (88% of 3,100 sheep and goats) graze south in the Moiynkum Desert or move seasonally to the north. Only 22% of the grazing (625 sheep and goats) is near the village. The pastoralists say there is little forage near the village. Furthermore, there are few opportunities for cash income. In Ulanbel, the intermediate case, village-based pastoralism is not possible but there are many other opportunities for cash income. There are about 10,000 sheep and goats in total. A few livestock (6%) are able to graze near the village for some of the year, but the majority graze further afield. Livestock have access to a series of Soviet era wells 30–50 km south of the village.

The main questions at this site are:

1. Is there any evidence of reduced NDVI or reduced RUE in grazing areas, particularly in village grazing areas, compared to other areas?
2. Have there been any trends in NDVI or RUE over the period 1989–2003?
3. Is there evidence of increased forage availability near the village of Sary Ozek, which would permit year around grazing by livestock?
4. Does low forage production near the village of Malye Kamkaly make village-based grazing unsustainable?
5. What are the spatial and temporal distributions of NDVI and green-up over the broader region of traditional long-range pastoral migrations?

Elevations are lowest in the Chu River valley with higher elevations to the north and south (Figure 8). The region is primarily desertic. A zone of tugais (floodplain forest), shrubs, and meadows occurs along the Chu River. Most of the area north of the river (Betpak-Dala) is vegetated with sagebrush (*Artemesia* spp.) and perennial saltwort shrubs (*Salsola* spp.) and desertic plains on fine textured soils. Scattered throughout are halophytic deserts (areas of high salinity) and solochak's depressions (depressions where salt accumulates on the soil surface). The southern portion, the Moiynkum Desert is a sandy desert comprised of plains and melkosopochnik (denuded hills and depressions). Vegetation is dominated by psammophytic (sand loving) shrubs (e.g. *Calligonum* spp. *Artemesia* spp.) and saxual (*Haloxylon* spp.) woodland.

Annual precipitation increases from approximately 155 mm in the west central portion of the study area to approximately 225 mm in the northeast and southeast (Figure 9). Over the broader area studied by Robinson and Milner-Guilland (2003), there is a low precipitation area centered just west of the Moiynkum study area, with increasing precipitation in all directions. The Moiynkum study area includes some of the driest areas of the larger Robinson study area. Rainfall has varied annually from a minimum of approximately 90 mm in 1995 to a maximum of approximately 320mm in 1993. Mean annual

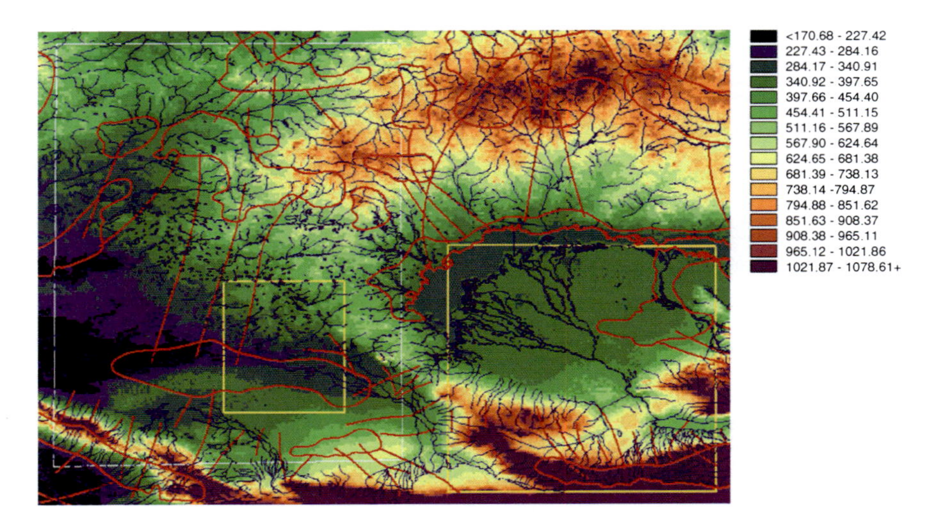

Figure 8. Elevation (meters), DARCA Kazakstan study regions (boundaries in yellow), Robinson and Milner-Guilland (2003) study region (boundary in cyan), the traditional seasonal migration areas and migration routes of M.G. Sakarov (Federovich 1973, Kerven et al. 2004) (in red), and hydrology (blue). Based upon a 2-minute resolution DEM of the world obtained from NOAA

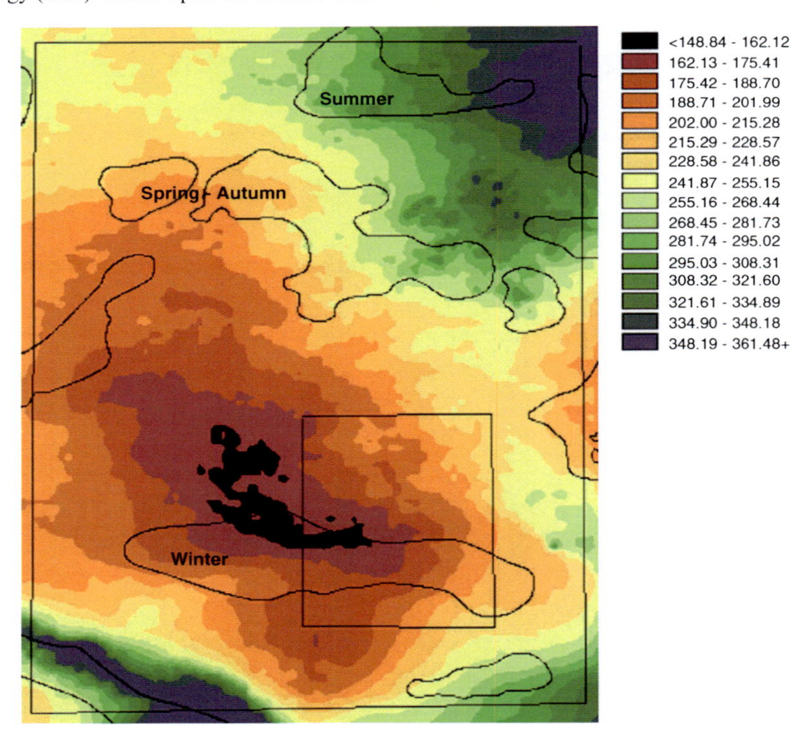

Figure 9. Mean annual precipitation (mm) across the Robinson and Milner-Guilland (2003) study area and the DARCA Moiynkum/Chu/Betpak Dala study area (smaller rectangle). Also shown are the historical pastoral grazing areas of M.G. Sakarov (Federovich 1973, Kerven et al. 2004). Seasonality of use is based upon Robinson and Milner-Guilland (2003)

precipitation was 170 mm over all. In the village grazing areas, mean annual precipitation was 185, 195 and 205 mm, going west to east. The CV's of the three areas were 0.34, 0.32, and 0.30.

NDVI generally decreases from south to north (Figure 10A), with much higher NDVI's in the tugai vegetation surrounding the Chu River. There is also a weak gradient from the southwest to the northeast. This pattern is similar to the pattern of topography (Figure 8), with higher elevations to the south and southwest. There is no apparent effect of increasing elevation to the north, however. The NDVI pattern was different from the precipitation pattern in that precipitation increased in the west and northwest, while NDVI did not.

The mean NDVI in Malye Kamkaly village grazing areas was not much greater than the mean NDVI values in the northern desert (Table 7). NDVI in the Sary Ozek area was considerably higher, while that of Ulanbel was intermediate. Higher values were due to the inclusion of greater portions of the floodplain. The amount of floodplain included in village grazing areas was highest in Sary Ozek, followed by Ulanbel, and then Malye Kamkale. There were positive trends in NDVI, in the Ulanbel and Malye Kamkaly areas. There was a stronger positive trend in NDVI at the Sary Ozek area. Stronger and more positive trends were likely due to changes in the level of water subsidization, since stronger trends occurred in areas with greater amounts of floodplain.

The spatial distribution of RUE indicated that the RUE of southern sandy desert was more productive for a given amount of rainfall than northern fine-textured desert (Figure 10B). Despite the northwest-southeast gradients in precipitation, there were no such patterns in RUE. There were no clear patterns of reduced RUE in the grazing areas compared to other areas nearby. RUE in remote grazing areas 1 and 2 were similar, although the vegetation species composition is different (Behnke and Temirbekov 2005). Vegetation in area 2 is considered to be overgrazed and weedy. There were also clear zones of altered reflectances around water points in this area in a Landsat scene from September 2003 (Behnke and Temirbekov 2005).

The larger study region examined by Robinson and Milner-Guilland (2003) covers the range of traditional long range pastoral movements. Across this broader region, it is clear that there is relatively more forage available north of the Moiynkum study area, where pastoralists historically moved in Spring through Autumn (Figure 11). This distribution of NDVI is generally in line with the distribution of precipitation (Figure 9). The area of high NDVI in the far northeast is also a high precipitation area. This is the fringe of a mountain range (Figure 2B). There is a low NDVI area to the west of the Moiynkum study area, where precipitation is low.

To understand the basis of the traditional large scale movements, it was informative to examine the timing of the onset of greenness along a north-south gradient. The broad scale pattern indicated early onset of greenness across

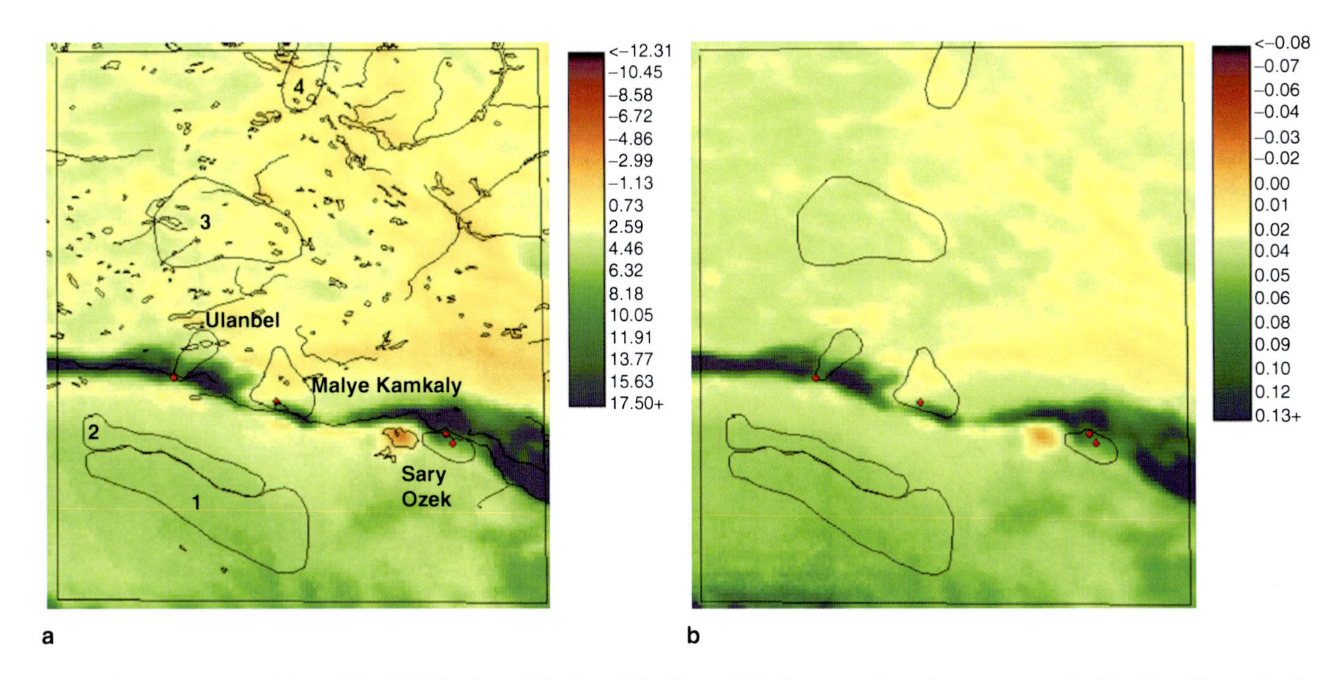

Figure 10. A) Mean annual NDVI 1991–2003 in the Moiynkum/Chu/Betpak Dala area. Also shown are roads (black), villages (red), remote grazing areas (1–4), and village grazing areas (black lines). B). Mean annual RUE 1991–2003 in the Moiynkum/Chu/Betpak Dala area

TABLE 7. Trend analysis and means of NDVI for Moiynkum/Chu/Betpak Dala grazing areas

Grazing Area	r2	P	Mean
Southern Remote - 1	0.22	0.145	7.30
Southern Remote - 2	0.29	0.088	7.91
Northern Remote - 3	0.21	0.153	4.96
Northern Remote - 4	0.22	0.147	5.56
Ulanbel	0.44	0.019	8.35
Malye Kamkaly	0.31	0.058	5.63
Sary Ozek	0.58	0.004	11.17

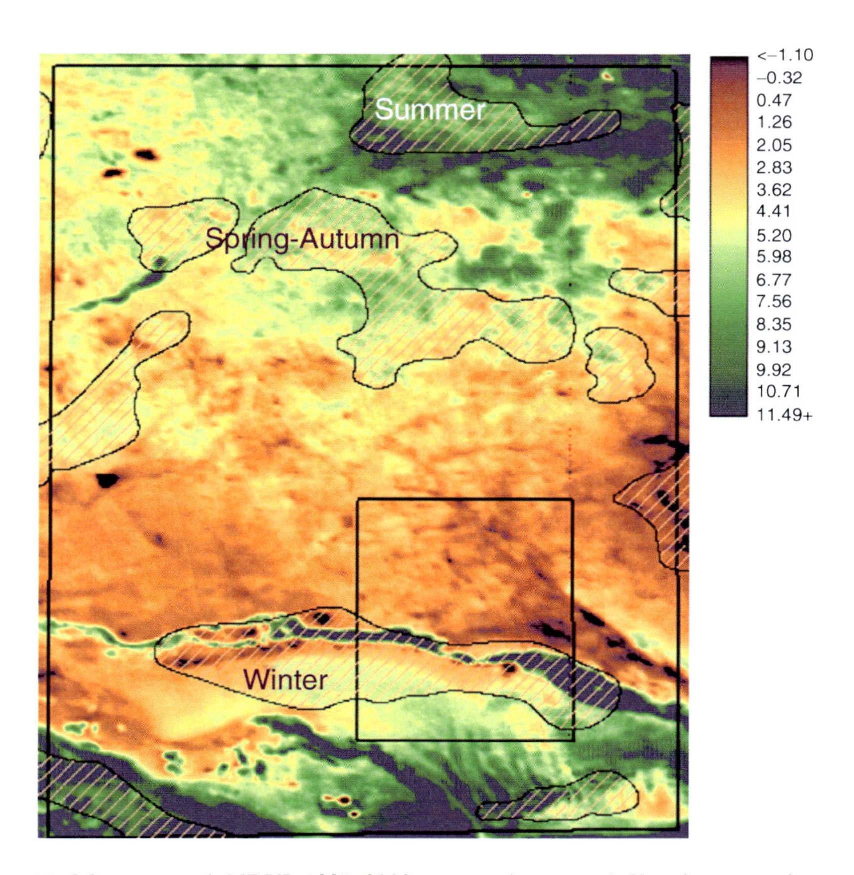

Figure 11. Mean annual NDVI 1991–2002 across the central Kazakstan study area of Robinson and Milner-Guilland (2003). Also shown is the Chu/Moiynkum/Betpak-Dala study area (smaller square), and the historical pastoral grazing areas identified by M.G. Sakarov (Federovich 1973, Kerven et al. 2004)

the south-central portion of the region (Coughenour 2005). To the north, the onset date was later, as it was to the south. Along two south-north transects, it could be seen that onset date declines from south to north until the Chu River. In the east, the onset date remained approximately level for some distance while in the west, the onset date began to advance. Along both transects, the onset date increased and reached a plateau in mid March in the north. This explains why summer pastures were located further north. The date of maximum NDVI exhibited a very weak pattern across the larger region. The date of maximum NDVI tended to be later in the northeast, and earlier in the southeast and across a band north of the Chu River. Elsewhere the date of maximum NDVI was sporadically distributed.

3.4. THE BALKASH BASIN/ALMATY OBLAST STUDY AREA, SOUTH-CENTRAL KAZAKSTAN

The easternmost Kazakstan study site covers a broad region from the mountain ranges in the south, north through the desert and into the Lake Balkash Basin. The traditional pastoral system involved large scale nomadic and transhumant movements along the north-south gradient with winters being spent in the areas that were less subject to cold and winds such as sandy dunes, spring grazing on the semi-desert, and summer grazing on the high meadows of mountains in the south and east or in the northern steppes, well north of the study area (Alimaev 2003). According to M.G. Sakarov, grazing in the Balkash area was mainly in the winter, with movements to the far north in summer (Kerven et al. 2004, Federovich 1973). Movements were compressed by the Russian administration in the late nineteenth century, followed by the formations of collective state farms in the 1930's. Long distance migrations run by the state were reinstated during the 1950's. During the Soviet era the state greatly intensified the system by adding some 150 new state sheep farms in previously little used areas, planting of improved forage species, fertilizer inputs, and mechanization to produce fodder. Since decollectivization private livestock holders cannot afford to produce such fodder. As a result, livestock populations plummeted, and potentially, rangelands have begun to recover from the intensive utilization during the Soviet period. (Alimaev 2003)

The primary questions for this study site are:

1. How does forage availability vary along a north-south transect across the region?
2. Is there evidence of rangeland recovery in terms of increased forage production and RUE?

This region supports a broad range of vegetation types, most likely due to the wide range of elevations (Figure 8) and resultant climatic conditions. Most of the northern, low-lying half of the study area is referred to by Alimaev (2003) as: "sands" or sandy deserts, or deserts, plains and melkosopochnik (Khramtsov and Rachkovskaya 2000). South of that, at low elevations, are the "plain desert" (Alimaev 2003) or desert, piedmont and low mountains (Khramtsov and Rachkovskaya 2000). "Semi-desert" (Alimaev 2003) is transitional between plain desert and steppe. Steppe vegetation occurs at intermediate elevations in the south. Subalpine vegetation is at higher elevations and is transitional between steppe and high mountain vegetation. Runoff from the mountains in the south feeds a river that flows northwards, fanning out in a deltaic system on the desert, and ultimately draining into Lake Balkash (Figure 8). Tugai, shrub, and meadow vegetation surround this system of rivers and tributaries. The vegetation is described in greater detail by Alimaev (2003).

Annual precipitation across the study area ranges from in excess of 950 mm at the highest elevations in the southeast to approximately 140–160 mm at low elevations in the northwest. The precipitation gradient is steep, reflective of the rapid change in elevation in the southern portion of the study area.

The southernmost sites at higher elevations received more moisture, reaching peaks of 650–750 mm in some years, while the driest sites received at most 200–250 mm per year and in dry years as little as 100–125 mm. CV's increased from 0.23 in the south to 0.35 in the north.. There were significant positive trends in annual precipitation over time in the southern portion of the study area (Coughenour 2005).

The spatial distribution of NDVI corresponded to the precipitation distribution, except for the water-subsided tugai and meadow vegetation zone (Figure 12), and the lower NDVI's at highest elevations that are due to cold temperatures and shorter growing seasons. Areas of exceptionally low NDVI (black areas) are water bodies, snow, or rock..

The spatial distribution of RUE was similar to the spatial distribution of NDVI (Figure 13). RUE was higher further south in areas with higher precipitation, except at high elevations. RUE showed a somewhat patchy distribution in the southern half of the study area, particularly in the semi-desert and steppe vegetation zones. Upon close inspection, it was clear that around some, but not all of the populated places there is lower RUE than further away. It cannot be ascertained what this is due to, but livestock grazing is one possibility.

There were positive trends in NDVI in most of the seven vegetation zones. Trends were highly significant in all vegetation zones except the northern sandy deserts and high elevation mountain vegetation (Coughenour 2005). However, there were no significant trends in RUE across the broad vegetation

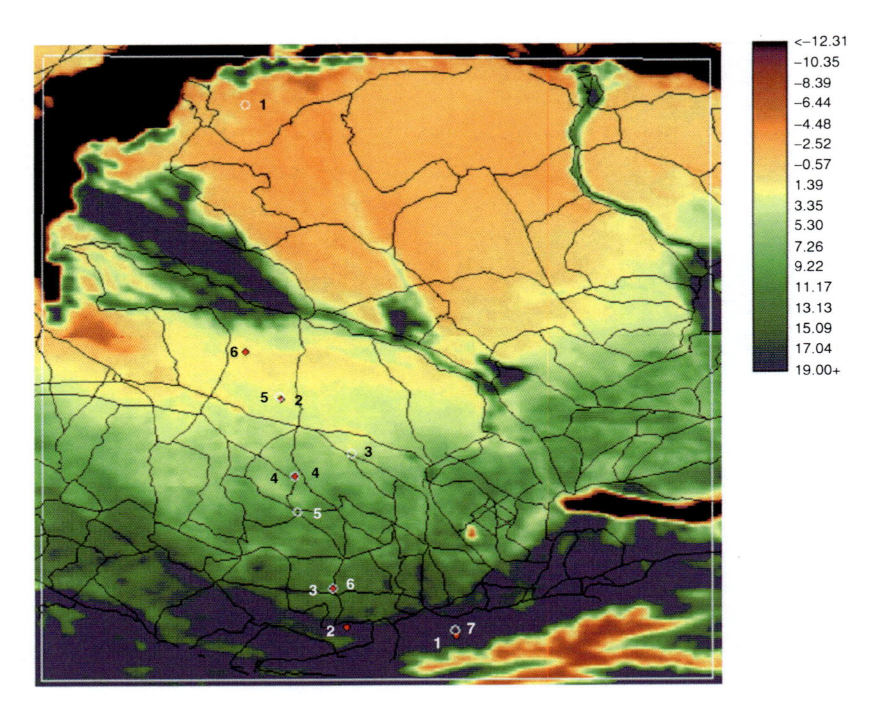

Figure 12. Mean annual NDVI 1991–2003 in the Balkash region, Kazakstan. Also shown are roads, the study areas of Ellis and Lee (2003) in red, and the study areas of Alimaev (2003) in white. Numbers on the right of the points are the Alimaev site numbers, numbers on the left are the Ellis and Lee site numbers

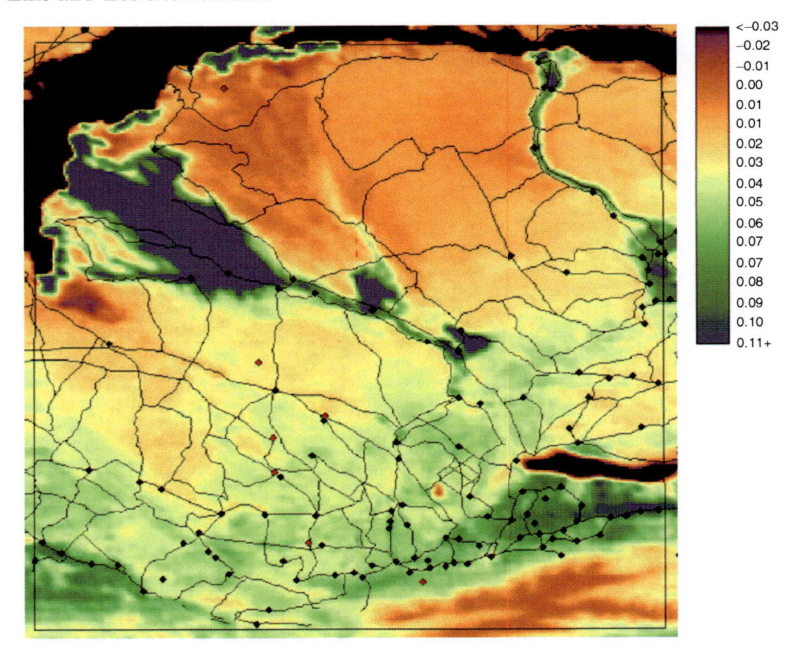

Figure 13. Mean annual RUE 1991–2003 in the Balkash region. Also shown are roads, populated places in black, and study sites of Alimaev (2003) in red

zones. The increases in NDVI at this scale could thus be attributed to increases in annual precipitation rather than an increase in basal plant cover.

The upwards trend in NDVI was also apparent at the study sites of Alimaev (2003) and Ellis and Lee (2003). In general, the more productive the site, the stronger the trend, with the southern sites exhibiting stronger trends than northern, more arid sites. The trends were highly significant ($p < 0.001$) at most sites (Table 9). At the driest sites, the trends were still significant ($p < 0.5$) however. In contrast, site RUE's showed fewer trends (Table 10). There was a significant upwards trend at sites 3, 4, and 6 of Alimaev. These sites are located in plains desert, semi-desert, or steppe.

The spatial distribution of changes in NDVI between 1991–1993 and 1999–2003, calculated as the difference in mean values of those two periods, showed general increases throughout the area with larger increases in the south (Figure 14). The magnitude of the increase is expected to be proportional to the mean NDVI, which increases to the south. As indicated above, the increase in NDVI was due largely to greater precipitation in the later period. However, RUE corrects for the difference in precipitation (Figure 15). The change in

TABLE 8. Trend analysis and means of RUE for Moiynkum/Chu/Betpak Dala grazing areas

Grazing Area	r2	P	Mean
Southern Remote - 1	0.09	0.333	0.043
Southern Remote - 2	0.02	0.675	0.040
Northern Remote - 3	0.08	0.382	0.026
Northern Remote - 4	0.08	0.374	0.024
Ulanbel	0.087	0.353	0.049
Malye Kamkaly	0.166	0.189	0.030
Sary Ozek	0.344	0.045	0.056

TABLE 9. Trend analysis and means of annual average NDVI*100 within a 15 km radius of the study sites of Alimaev (2003). Slope is change in NDVI*100 per year. Sites are arrayed north to south

Site	r2	P	Mean
1	0.48	0.14	2.0
2	0.61	0.048	6.5
3	0.82	0.002	9.5
4	0.87	<0.001	6.5
5	0.89	0.001	14.0
6	0.73	<0.001	17.6
7	0.87	0.001	28.8

TABLE 10. Trend analysis and means of annual average RUE (NDVI/mm) within a 15 km radius of the study sites of Alimaev (2003). Slope is change in RUE per year. Sites are arrayed north to south

Site	r2	P	Mean
1	0.38	0.24	0.010
2	0.16	0.64	0.035
3	0.82	0.002	0.035
4	0.62	0.04	0.038
5	0.36	0.28	0.043
6	0.40	0.02	0.044
7	0.22	0.50	0.045

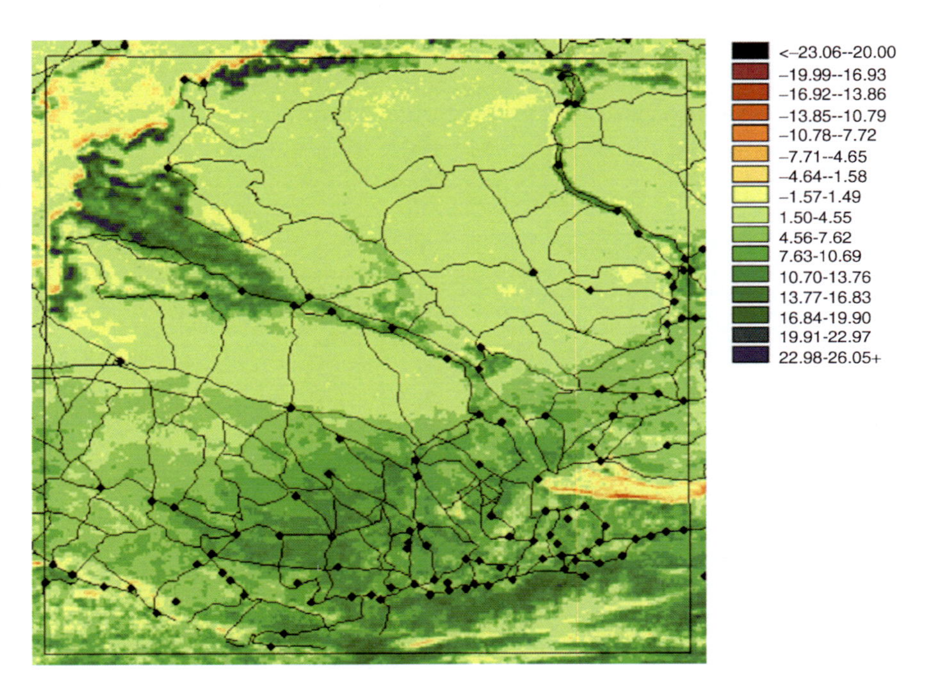

Figure 14. Difference in mean NDVI in the Balkash region between 1991–1993 and 1999–2003 (positive indicates increases). Also shown are roads and villages

RUE was less pronounced, but there were widespread areas where RUE clearly increased (light green and dark green areas). In other areas there was little change (yellow areas). Interestingly, the areas where there were increases corresponded to denser concentrations of population places (villages) and roads. This likely reflects rangeland recovery in those areas that were more likely to have experienced heavy livestock grazing in the Soviet era.

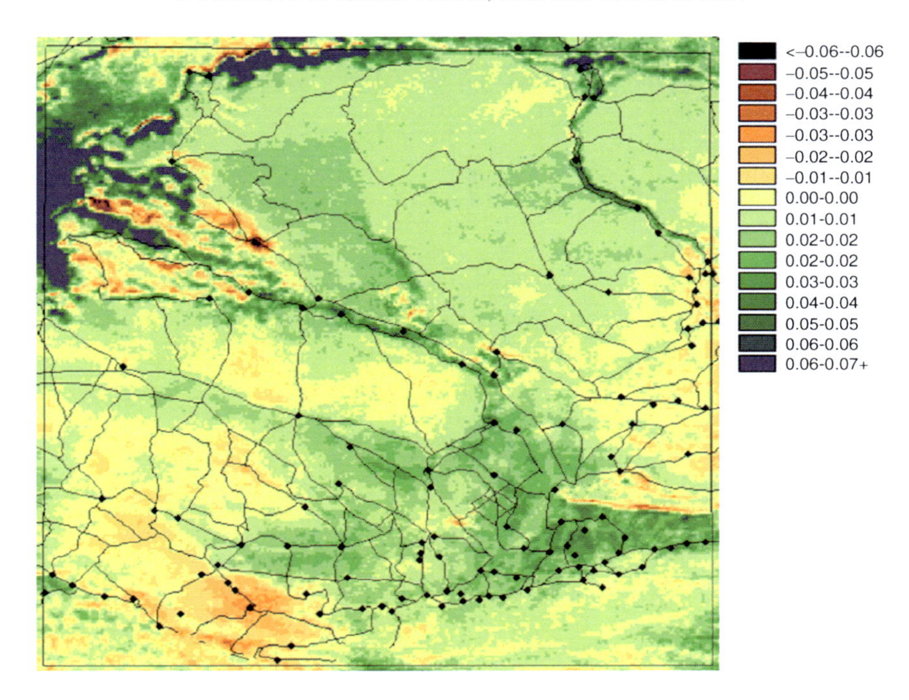

Figure 15. Difference in mean RUE in the Balkash region between 1991–1993 and 1999–2003 (positive indicates increases). Also shown are roads and villages

4. Discussion

4.1. GOKDEPE

Forage biomass varied seasonally along the north-south gradient of livestock movement, however all of the village areas exhibited similar seasonal NDVI patterns, with respect to the time of green-up, and time of maximum NDVI. Thus, north-south movements cannot be driven by vegetation phenology. This is in agreement with the finding that north-south movements are driven mainly by water quality, with use of fresh water sources in the summer in the south and more saline sources in winter in the north (Behnke et al. this volume).

Sedentarization has probably led to some degree of range degradation near some, but not all villages. NDVI and RUE was higher outside than inside the northern four village grazing areas. On average, there was a gradient of decreasing NDVI and RUE with decreasing distance to individual villages. Furthermore, there was visibly lower RUE around some, but not all of the villages (Figure 3A). These results are in agreement with the findings of Khanchaev (2005), who found about half as much forage biomass 1 km from villages wells as 5 km from village wells. He also found normal species composition 5 km from village wells but a degraded species composition 1 km from village wells. Although he found evidence of degradation 1 km from winter wells, the NDVI/RUE data did not detect differences between 0–3 km and 3–5 km from winter wells.

The gradient of decreasing NDVI and particularly RUE from the far north to the south (Figure 3) coincides with the lack of water in the north, and decreasing distance to water towards the south. This may not be coincidental. Khanchaev (2005) noted that the productivity of an important forage species, *Carex physodes*, declines to the south and is eventually replaced with annuals. This species also declined within 1 km of wells compared to 5 km from wells. Thus it is possible that at a larger scale, grazing in the Gokdepe area has induced species shifts and reduced productivity. Further research would be needed to confirm this suspicion.

There was no strong time trend one way or the other in rangeland condition at this site. Although NDVI did show an increasing trend, RUE did not show any clear trends in any of the village grazing areas or within 10 km of any type of well. The increasing trend in NDVI was therefore due mainly to an increase in precipitation, rather than fundamental rangeland recovery. There were weak positive trends in RUE in some locations, suggesting improved rangeland condition (Table 4). Within 10 km of winter wells, P values for trends were less than or equal to 0.14. At 5–10 km from summer wells and villages P values were 0.16. The results are suggestive, but a longer time series will be necessary to firm up any conclusion that range conditions are improving.

The results indicate that the pastoral system is generally sustainable, and that livestock movement is likely key to this sustainability. There is very localized degradation near settlements, but otherwise livestock grazing pressure appears to be relatively well distributed. Water-driven movements between summer wells and winter wells have likely contributed to improved livestock distributions, in the sense that livestock cannot use winter wells in summer, thus reducing grazing pressure there to some extent. Similarly, movements to utilize forage near winter wells in winter reduces grazing pressure in the summer well areas. Livestock numbers also appear to be distributed in proportion to forage productivity (Behnke et al. this volume), implying that grazing pressure is also well distributed rather than being concentrated at unsustainable densities in certain areas. Nevertheless, at a larger scale, grazing may be influencing rangeland condition.

4.2. RAVNINA

There were strong seasonal variations in green forage availabilities in the three grazing areas (village, winter, summer), with greenup occurring in early February in village areas and late February in summer areas. The difference in onset of green-up dates between winter and summer use areas suggests that the timing of usage of these areas is logical, because green-up in summer areas occurs later.

Summer pastures are likely to be better forage sources in summer. They are not much better than the winter pastures in April, and if anything they have equal or less green biomass than winter pastures in June through August (Figure 5). However, while green biomass may be comparable in summer on winter vs. summer use areas, the total amount of biomass on summer pastures is likely to have been greater, due to the higher NDVI's earlier in the growing season. Mean annual NDVI, equivalent to the time integral of NDVI, was also higher in summer grazing areas than winter grazing areas (Table 5). It must be remembered that NDVI senses only green biomass, rather than total biomass. Green biomass produced earlier in the season was likely converted to senescent biomass during summer, and this biomass would not have been sensed by the NDVI. Therefore, the reason for using summer pastures in the summer may due partly to the later onset of green-up further south, and partly due to the greater quantity of forage produced in those areas. Other factors such as colder weather in the south may have also made those areas less desirable to use during winter

Because forage production is likely higher in the summer ranges in the south, private flocks kept north of the canal are likely at a disadvantage with respect to forage. NDVI values were highest in the summer use areas, followed by the winter use area, and after that by the village areas. The implications are that privatized livestock herds restricted to the village areas are indeed at a disadvantage with respect to foraging conditions, which could potentially lead to lower rates of intake and reduced weight gain. An important advantage of moving to the summer range, however, is to access far more land area, and thus far more forage in total. Thus, livestock managed by grazing associations south of the canal have greater forage production available to them. The higher NDVI values further south were likely mainly due to higher precipitation amounts at higher elevations. However, RUE's of summer grazing areas in the south were also higher than RUE's in the village grazing area (Table 5).

There did not appear to be any evidence of change in range condition based upon NDVI or RUE over the 1989–2003 time period. There was an increase in NDVI between 1989 and 1994 due both to increases in RUE and precipitation. However, the droughts of 1995 and 2001 reversed these increases. As a result there were no consistent trends in NDVI or RUE over the whole time period except in meadows along the canal, which could have been due to increases in irrigated crops or flows in the canal. Comparisons of NDVI and RUE inside and outside the village grazing area and the summer grazing area also showed no evidence of degradation or reduced production due to grazing. The only area where there might have been reduced RUE was along the southern border of the canal in the west. It is possible this reduction could have been due to greater livestock grazing pressure close to the canal. However, other explanations cannot be ruled out without further study of possible soil and vegetation differences.

Thus, on the whole, it appears that pastoralists in this area have done well in distributing their livestock throughout the area, and livestock densities appear to have been kept to sustainable levels in terms of rangeland condition. The formation of grazing associations which share common grazing resources at distance from villages has apparently been a successful solution to resource utilization in this area. Notably, it has not led to a "tragedy of the commons". In contrast, flocks owned by individual herders kept within the village area are likely at a disadvantage due to lack of access to a more expansive and richer forage base in the south. The associations in effect, enjoy an economy of scale, in that access to a larger and more varied pool of resources can more efficiently be used by larger collective herds. More herders likely provide an increased range of options for grazing, and greater adaptability. It cannot be determined here if an individualized, but more expansive and mobile system may be a viable alternative to the associations. However, it would be a complex and unwieldy problem to try to allocate specific small-scale grazing areas to individual flocks.

4.3. MOIYNKUM/CHU/BETPAK-DALA

There was no substantial evidence of range degradation in the form of reduced NDVI's or RUE's in any of the grazing areas, in comparison the mean NDVI's or RUE's of the vegetation zones as a whole. Increases in NDVI in response to higher precipitation after 1996 occurred in grazing areas as well as entire vegetation zones. This suggests that stock numbers are limited by forage to levels that do not reduce plant productivity. However, the NDVI and RUE do not sense changes in plant species composition or forage quality. Evidence from ground observations and satellite-based vegetation classification have shown that species composition is different in remote grazing area 2, where there are numerous wells and thus historically high livestock densities. Shifts in species composition may also have occurred some time ago, for example during a period when livestock were subsidized with fodder.

Although there were significant increasing trends in NDVI in the desert vegetation zones, there were no significant trends in RUE. Therefore the NDVI trends were due to increased precipitation. NDVI and RUE both increased in the floodplain (tugai) vegetation along the river, possibly reflecting increases in irrigation or increased flows in the river. These increases would provide increased forage in village grazing areas, particularly in villages with more extensive floodplain availability. Results showed that NDVI and thus forage availability are strongly linked to precipitation (Coughenour 2005). Drought years are an inherent feature of arid environments, consequently it would be important to find ways to ameliorate drought when it does occur. One possibility might be to make increased use of floodplain vegetation at a broader scale to create a drought fodder reserve.

Differences in NDVI and RUE among the three villages potentially explain their divergent histories. It was clear the forage availability would be much lower in Malye Kamkaly than the other two villages. This would explain why village-based pastoralism is untenable there. NDVI data suggested that limited access to floodplain vegetation was likely the primary cause. Not only does floodplain access improve forage availability in general, it also provides a buffer in drought years such as in 1995. In 1995 the NDVI in Malye Kamkaly dropped to very low levels, which would have severely limited livestock, resulting in mortality or outmigration to other areas. The few remaining pastoralists there only survive through seasonal movements to distant pastures in the desert (Kerven et al. 2004). In contrast, Sary Ozek is highly suited for village-based pastoralism due to high access to floodplain vegetation (Kerven et al. 2004). This undoubtably explains why a high percentage of livestock graze in the village grazing area year around. Furthermore, NDVI and thus forage, did not drop to extremely low values in the drought year of 1995 so there would have been more surviving livestock. Despite the possibility for village-based pastoralism in Sary Ozek, the total number of livestock based there is small compared to Ulanbel (Behnke and Temirbekov 2005). There, the majority of the livestock utilize remote grazing areas which are much more extensive, therefore containing a greater total quantity of forage. In particular, access to the southern grazing areas where there are wells is probably a key reason why many more livestock can be supported. At the same time, a sufficient degree of village-based pastoralism is possible due to access to floodplain vegetation. The village-based livestock likely provide an important supplementary source of income for village families.

Movements to remote grazing areas provide increased forage availability simply because a greater area of land is utilized. A wider range of forage resources is also to be expected, given the diversity of plant communities throughout the region. Although no temperature data were available, it is likely that it is significantly warmer in the south than the north during winter, and this could be a reason for utilizing northern pastures in summer. Although forage production is lower in the north, it is still a worthy resource if it can be utilized under suitable environmental conditions.

However, at the scale of movements exhibited currently in the study area, there are relatively limited options for moving in response to seasonal forage availability. A precipitation gradient exists, but it is a small gradient completely within the desertic range. Seasonality of vegetation is likely to be primarily determined by temperature rather than precipitation within the study area. North-south movements could be timed in response to date of onset of greenness, but there was no north-south gradient in the time of maximum NDVI. A gradual movement of livestock northwards beginning from the Chu River would follow the progressive later dates of green-up. Vegetation that is just greening up is likely to have higher protein and digestible energy content, so forage quality could be maximized by such a movement.

A richer range of opportunities for movement would have existed in the broader scale traditional pastoral range. Across the broader region, the onset of green-up is about 30 days later than the furthest north areas of the smaller study area. The north-south range of temperatures would be considerably greater. More importantly perhaps, is the fact that traditional summer pastures in the far north receive significantly higher precipitation (300–350 mm/year, see Figure 9). It would be logical for pastoralists to have moved northwards in the spring as the green-up advances. Perhaps more important reasons for them to have moved north would have been to take advantage of the higher productivity in those areas (NDVI was higher further north, see Figure 12), and simply to be able to utilize the greater quantity of forage afforded by utilization of a greater land area. While livestock were migrating to the north, the pastures in the southern deserts would have been ungrazed, leaving a more ample forage base for winter grazing at warmer temperatures. The contracted migration pattern that we see today omits the former transitional and summer pastures areas to the north, thus resulting in considerably less forage available to livestock. An interesting question to ask is whether the total number of more sedentarized and less mobile livestock supported throughout the broad region is now greater than the number that could be supported under the traditional system of long-range mobility?

4.4. BALKASH/ALMATY OBLAST

Rangeland recovery in terms of increased forage production and RUE might be the expected response to the massive destocking that occurred in the early 1990's. There was evidence that forage production in 1998–1999 was greater in steppe, semi-desert and plain desert than in 1974–1986 (Alimaev 2003). The NDVI data here do not permit comparisons to conditions prior to 1991 in this area. However, given that the destocking occurred through the early 1990's, trends could still be expected 1991–2003. Indeed, there were very significant positive trends in NDVI in all vegetation zones except the northern sandy desert and the alpine (Coughenour 2005). However, a large part of this increase was due to an increasing trend in precipitation. Positive precipitation trends were significant at the semi-desert, steppe, and alpine sites studied by Alimaev, but not in the sandy desert or plain desert.

A more fundamental measure of range condition is the RUE, since it corrects for precipitation differences. The map of RUE over the region (Figure 13) showed a patchy distribution in the south central part of the study area. Averaged over the entire vegetation zones, there were no trends in RUE. However, there were significant upward trends in RUE at 3 of the sites studied by Alimaev (2003) and Ellis and Lee (2003). These sites were located in the plains desert, semi-desert and the steppe. There were no trends at sandy desert or subalpine sites. The map of RUE change between 1991–1993 and 1999–2003 showed considerable areas of increased RUE as well as considerable areas of no change. These did not correspond to particular vegetation types. Instead, the areas of increases were largely in areas where there were populated places and denser road networks. This is the pattern that might be expected if rangeland degradation occurred primarily on the former collective farms, or near the mechanized feeding stations, watering points, and settlements. Therefore, RUE and range condition has probably increased somewhat over this time period in areas that were formerly more heavily grazed, irrespective of vegetation type.

The traditional pastoral system in this region involved long distance movements from winter use areas in the sandy deserts of the north to summer use areas in the steppes and mountain meadows in the south, as well as east of the study area. Pastoralists were first settled into collectives beginning in the 1850's, but long distance moves were reintroduced in the 1930's, (Alimaev 2003). This seasonal use pattern was quite productive for a period of time, until sedentary sheep farms were established in the semi-deserts and deserts where there was previously only seasonal use in the winter (Alimaev 2003). The traditional movement system would have allowed regrowth in the drier portions of the region during the growing season with little or no grazing pressure. Sedentarized pastoralism in contrast, does not. The provision of fodder and other subsidies from outside local areas would have supported livestock densities that are above the levels that could be sustainably supported by the local pastures. With the end of these subsidies and resultant destocking, pasture condition may recover, as discussed above, but the question is: Would it be more sustainable as well as productive to reinstate the traditional movement system? Mean NDVI in the deserts was very low, indicating that only very low livestock densities could be supported, and the livestock may be relatively unproductive. In contrast, NDVI in the steppes, mountain meadows and alpine was 2–4 times higher. Alimaev (2003) similarly found that forage production in the steppe was roughly two times greater than in the semi-desert and forage production in the sandy desert was roughly 80% of that in steppe. Thus, by moving south and to higher elevations in summer, livestock could gain weight much more rapidly and build up body reserves that could carry them through the winter in the less productive but warmer deserts. Thus, with respect to rangeland sustainability as well as livestock productivity, the traditional system was very rational.

5. Conclusions

Each of the four case studies revealed a different situation with respect to the distribution of resources, the importance of movement, the degree to which movements have been altered, and the consequences for rangeland condition. The analyses revealed spatio-temporal patterns of precipitation, forage, water, and topography which necessitate movement and adaptability. Although traditional regional scale migrations have been lost, the analyses suggested that smaller scale movements, coupled with appropriate stocking rates, can partially avert the negative consequences of complete sedentarization for pastoral production and rangeland condition. North-south movements in the BalkashAlmaty region of Kazakstan were driven by climatic gradients of precipitation and winter weather. Access to the more productive rangelands in the south in summer and to pastures at lower and warmer elevations in winter is vital. Sedentarization of pastoralists onto collective farms coupled with fodder and other subsidies contributed to rangeland degradation. However, since the end of subsidies, the decline in livestock numbers have resulted in rangeland regeneration. Similarly in the MoiynkumChuBetpak-Dala region, traditional movements between higher, more productive rangelands in summer and lower elevation pastures have been mostly abandoned. Smaller scale movements to pastures located at distance from villages is sustaining the pastoral system, with some sacrifice in range condition around water points. Village based pastoralism has proven to be viable only where there is ample access to floodplains along the Chu River. However these systems, while viable, are likely to be marginally productive as they do not make use of the more productive high elevation pastures used previously. Reestablishing these movements would probably lead to increased productivity and sustainability. Smaller scale movements also proved to be important in Turkmenistan. At the Gokdepe site, livestock movements driven by water as well as forage distribution are important to the sustainability of the pastoral system as a whole. Increased sedentarization in villages has led to localized degradation at Gokdepe. Localized degradation around water points may be inevitable. However, seasonal, albeit smaller scale, movements driven by water as much as forage, have largely distributed grazing in proportion to resources, at least partially averting further degradation. At the Ravnina site, good mobility is achieved through the formation of grazing associations which utilize common pastures remote from villages. Village-based pastoralists in Ravnina are at a disadvantage compared to more mobile herders. Thus, to maintain or improve pastoral productivity and sustainability in Central Asia, it will be essential to integrate the spatial distributions of forage, water, climate, and the effects of alternative livestock movements on the condition of the livestock as well as the condition of the rangelands.

Acknowledgement

This research was funded by a grant from the European Union to the project "Desertification and Regeneration: Modeling the Impact of Market Reforms on Central Asian Rangeland. in Central Asia" (DARCA). We are indebted to the late James E. Ellis, of Colorado State University, for envisioning this research.

References

Alimaev, I. I., 2003, Transhumant ecosystems: Fluctuations in seasonal pasture productivity, in: *Prospects for Pastoralism in Kazakstan and Turkmenistan*, C. Kerven, ed., Routledge Curzon, London, pp. 31–51.

Behnke, R. H., 2003, Reconfiguring property rights and land use, in: *Prospects for Pastoralism in Kazakstan and Turkmenistan*, C. Kerven, ed., Routledge Curzon, London, pp. 75–107.

Behnke, R. H., and Temirbekov, S., 2005, Desertification and Regeneration in Central Asia, DARCA Project, Report to the European Union, MacCauley Institute, Aberdeen.

Coughenour, M. B., 1992, Spatial modeling and landscape characterization of an African pastoral ecosystem: a prototype model and its potential use for monitoring drought, in: *Ecological Indicators*, D. H. McKenzie, D. E. Hyatt and V. J. McDonalds (eds.). Vol. I. Elsevier Applied Science, London and New York, pp. 787–810.

Coughenour, M., 2005, A Remote Sensing Analysis of Trends in Rangeland. Productivity and Condition in Central Asia, Report to DARCA Project, Natural Resources Ecology Laboratory, Colorado State University, Fort Collins, Colorado.

Ellis, J., and Lee, R., 2003, Collapse of the Kazakstan livestock sector: A catastrophic convergence of ecological degradation, economic transition and climatic change, in: *Prospects for Pastoralism in Kazakstan and Turkmenistan*, C. Kerven, ed., Routledge Curzon, London, pp. 52–74.

Federovich, B. A., 1973, Natural conditions of arid zones of USSR and the ways in which livestock husbandry has developed in them, Institute of Ethnography History of the economy of the peoples of Central Asia and Kazakhstan, Leningrad, Nauka, **XCVIII**, pp. 207–222.

Justice, C., Hall, D., Salomonson, V., et al., 1998, The Moderate Resolution Imaging Spectroradiometer (MODIS): land remote sensing for global change research, *IEEE Transactions on Geoscience and Remote Sensing* **36**: 1228–1249.

Kerven, C., 2003, We have seen two worlds' impacts of privatization on people, land and livestock in: *Prospects for Pastoralism in Kazakstan and Turkmenistan*, C. Kerven, ed., Routledge Curzon, London, pp. 1–9.

Kerven, C., Alimaev, I. I., Behnke, R., Davidson, G., Franchois, L., Malmakov, N., Mathijs, E., Smailov, A., Temirbekov, S., and Wright, I., 2004, Retraction and expansion of flock mobility in Central Asia: costs and consequences, *African Journal of Range and Forage Science,* **21**: 159–169.

Khanchaev, H., 2005, Result of monitoring the vegetation in Gokdepe pastures, Report to DARCA Project.

Khanchaev, K., Kerven, C., Wright, I. A., 2003, The limits of the land: pasture and water conditions, in: *Prospects for Pastoralism in Kazakstan and Turkmenistan,* C. Kerven, ed., Routledge Curzon, London, pp. 194–209.

Khramtsov, V. N., and Rachkovskaya, E. I., 2000, *Vegetation of Kazakhstan and Middle Asia (desert region)*, Department of Vegetation Geography and Cartography, Komarov Botanical Institute, Russian Academy of Sciences. St. Petersburg.

Lunch, C., 2003, Shepherds and the state; effects of decollectiviation on livestock management, in: *Prospects for Pastoralism in Kazakstan and Turkmenistan,* C. Kerven, ed., Routledge Curzon, London, pp. 171–193.

Prince, D. D., de Colstoun, E. B., and Kravitz, L. L., 1998, Evidence from rain-use efficiencies does not indicate extensive Sahelian desertification, *Global Change Biology,* **4**: 359–374.

Robinson, S., and Milner-Guilland, K. J., 2003, Contraction in livestock mobility resulting from state farm reorganization, in: *Prospects for Pastoralism in Kazakstan and Turkmenistan,* C. Kerven, ed., Routledge Curzon, London, pp. 128–145.

Tucker, C. J., and Sellers, P., 1986, Satellite remote sensing of primary production, *International Journal of Remote Sensing,* **7**: 1395–1416.

CHAPTER 5

THE IMPACT OF LIVESTOCK GRAZING ON SOILS AND VEGETATION AROUND SETTLEMENTS IN SOUTHEAST KAZAKHSTAN

SOUTH KAZAKHSTAN PASTURE USE RESULTS

ILYA I. ALIMAEV[1], CAROL KERVEN[*2], AIBYN TOREKHANOV[1], ROY BEHNKE[2], KAZBEK SMAILOV[1], VLADIMIR YURCHENKO[1], ZHEKSINBAI SISATOV[1], AND KANAT SHANBAEV[1]

[1] *Kazakh Scientific Centre for Livestock and Veterinary Research, Dzandosov Str. 31, 480035 Almaty, Kazakhstan*
[2] *Macaulay Institute, Craigiebuckler, Aberdeen AB15 8QH*

Abstract: This paper documents the impact of grazing on pastures and soil composition at study sites in southern and central Kazakhstan following decollectivization. Uncontrolled grazing and high stocking rates around settlements have produced both environmental degradation and diminished livestock performance despite overall declines in sheep numbers in the post-socialist period.

Keywords: Rangeland degradation, livestock mobility, pasture condition, Kazakhstan

1. Introduction

Livestock numbers have fluctuated widely in Kazakhstan from 1916 (when records are first available) to 2003. Over this period there were years with very high livestock populations (1928, 1976, 1981) combined with years in which populations were very small (1921, 1934, 1997–2000). The last crash in livestock numbers took place in the mid-1990s following decollectivization and continued up to 1999 when animal numbers stabilized. At present almost all livestock in Kazakhstan are owned privately, often by rural people who own very few animals. These small-scale livestock owners are forced to graze

* To whom correspondence should be addressed. Carol Kerven, 2 The Ridgeway, Great Wolford, Warwickshire, CV36 5NN, UK; e-mail: carol_kerven@msn.com

R. Behnke (ed.), The Socio-Economic Causes and Consequences of Desertification in Central Asia, 81–112.

their animals within a 5 km radius from their village due to socio-economical reasons. Nowadays villages can contain up to a hundred households each of which has livestock that are grazed year-round on the same kind of pastures adjacent to the village, causing a serious decease of pasture productivity. This situation also restricts the ability of the farmer to increase his herd size or improve the quality of the products produced by the herd.

Lack of a planned approach to pasture use may eventually lead to severe pasture degradation which will inevitably cause a decease in livestock numbers. The solution recommended in this paper is to apply a spatial approach to pasture use and to use the fodder potential of regenerated pastures that may be distant from settlements. As has been demonstrated by the analysis of ninety years of experience in Kazakhstan before, during and after the Soviet period, this kind of mobility and dispersal improves the pasture use system and sustains increased numbers of livestock.

The problem of pasture use in Kazakhstan therefore requires a scientific approach with detailed studies of the negative impact of the fragmented use of pastures combined with a clear explanation of environmentally safe ways for raising animals on pastures. The research summarized in this paper was based on a combined examination of soils, vegetation, and livestock productivity in an attempt to assemble the main elements of a comprehensive assessment.

Research was carried out at study sites in the Alatau-Balkhash hinterlands, in the Moinkum-Betpakdalla deserts and Ili-Jungar pasture complexes in Almaty and Jambul Oblasts of Kazakhstan. In Almaty Oblast the research work was undertaken along a north-south transect that stretched through Alatau-Balkhash pasture complexes from the Alatau Mountains to the Taukum Sands with total length of 350 km. Transects started in the south in the mountains and crossed steppes, semi-steppe, semi-desert, desert zones, following a traditional nomadic migratory route. Three villages along the transect were selected for intensive study: Shien (Plate 1a, 1b), Ulguli (Plate 2a) and Aydarly (Plates 2b, 3a, 3b).

A second north-south transect was located in Jambul Oblast, and stretched from the Moinkum Sands in the south (Plate 4a), through the Chu River valley and clay soil Betpakdalla pastures (Plate 4b), and finished in the north at the southern edge of the Sary Arka steppe. This transect provides a uniquely favorable environment for mobile livestock management. This huge territory covers 600 km from north to south and total area makes 10 million ha. We selected 11 vegetation sampling sites on different pasture types along this transect over a distance of 220 km. During the years of research work from 1999 to 2004 there were significant variations in weather conditions from year to year. The years 2002–2003 were dry, rainfall was average in 2001 and 2004, and 1999–2000 period was wet.

Plate 1a: Mountain pastures at 1,100 meters altitude. Summer grazing for the village of Shien, Almaty Province. Photograph by Iliya Alimaev
Plate 1b: Goats on foothill pastures in the vicinity of Shien. Photograph by Iliya Alimaev

Plate 2a: Cattle grazing on poor winter pastures within five km of the village of Ulgule, Almaty Province. Photograph by Iliya Alimaev
Plate 2b: Sampling regenerated *Artemisia* ephemeron pastures in spring. Photograph by Iliya Alimaev

Plate 3a: Weighing samples of winter pasture near the village of Aydarly, Almaty Province. Photograph by Iliya Alimaev
Plate 3b: Sheep grazing on winter pastures near Aydarly. Photograph by Iliya Alimaev

a

b

Plate 4a: Sheep grazing in the Moinkum desert, Jambul Province. Photograph by Iliya Alimaev
Plate 4b: Taking soil samples in the Betpakdalla, Jambul Province. Photograph by Iliya Alimaev

Soil types in the different zones include sierozem (grey desert soil), mixed with sand, pure sand and in some places with solid particles of clay. There are extensive areas of sierozems that characteristically contain little humus or carbonates. The other basic soil type of soil represented in study sites is chestnut soil.

2. Grazing impact

Grazing has a significant impact on vegetation causing many morphological and physiological changes, affecting both the growth processes of plants and the species composition of vegetation communities in pastures. First of all, grazing disrupts the accumulation and distribution of nutritious elements within plants. Vegetation influenced by grazing also undergoes morphological changes both at root level and above ground. Generally there is a change in the size of plant and quantity of leaves, quantity of flowers at blooming period, fruit and seed production, percentage of green and woody parts, percentage of sprouts with leaves and reproductive sprouts. Changes in the root system typically leads to relocation of the roots closer to the surface and as a result roots can not access water sources located deep in the ground, which may be necessary in a dry climate.

Consequences of the intensive grazing on pastures could be observed on the growth of *Artemisia terrae albae* at the research sites. We could also see quite an obvious change in vegetative cover (Table 1).

This table shows the accelerated growth of *Artemisia* in terms of bud formation, flowering and the formation of seeds on intensively grazed

TABLE 1. Grazing impact on *Artemisia terrae-albae's* phenological phases (Malae Kamkaly village, Jambul Oblast, average for studied years)

	Grazing		
Growth phase	Very intensive	Intensive	Weak
Regenerated vegetation	7.1; V[1]	30; III	30; III
Growth of sprouts	Up to 2.0; V	12; IV	9; IV
Bud formation	None	10; IX	Up to 15; IX
Flowering	None	Up to 10; X	Up to 5; X
Seed formation	None	Up to 20; X	Up to 25; X

[1]Arabic numbers refer to thousands of observations (plants, sprouts, buds, flowers or seeds) per hectare; Roman numerals refer to the months in which a growth process was observed.

pastures. However, at very high levels of overgrazing the plant becomes unable to complete all phases of its growth cycle.

Table 2 shows the impact of intensive grazing on plant numbers, even forage species that are resistant to trampling and grazing such as *Salsola oreintalis,* a perennial small shrub found commonly in the Betpakdalla and which is a good source of nitrogen. Compared to lightly grazed plant populations, plant numbers under intensive grazing were reduced to a third of their former level. At the same time 73% of plants in intensively grazed areas were old and only 16% were comparatively young. The most positive ratio of plant age groups could be found under light grazing conditions, while intensive pasture use led to a decrease in young plants and increasing numbers of unproductive and woody plants.

According to Zhambakin (1995) seed formation is the main indicator for identifying the viability of pasture plants. Active seed formation is not possible with intensive grazing, as demonstrated in the results of our studies summarized in Table 3. Strong reproductive growth of *Artemisia lessingiana* occurs only in favorable years that combine warm weather and high precipitation. In recent years, 2002 provided the most favorable conditions. Analysis of plant reproduction and development between 2000 and 2002 revealed that only half of potentially viable plants survived and that most of the vegetation died from trampling by animals.

The negative impact of grazing on the development of plant root systems has been studied on *Aremisia*-ephemeron pastures around Aydarly. Research work results conducted by the method of Larin (1956) provided us with the following data (Table 4). This table demonstrates the sensitivity of the root development process to variations in the intensity of gazing.

Table 5 shows how the dominant shrub species in *Artemisia*-ephemeron pastures is affected by different levels of grazing intensity and moisture. The

TABLE 2. Age variation of *Salsola oreintalis* depending on the intensity of grazing (Betpak dalla, average data for studied years)

Grazing intensity	Plant age categories				
	Young	Middle aged	Old	Very old	Total.
Weak	16.3[1]	18.5	8.0	6.6	49.8
	29.6[2]	37.1	16.0	13.4	100
Intensive	16.7	22.0	9.7	7.0	55.0
	33.5	40.0	17.6	12.8	100
Very intensive	1.1	1.3	1.5	10.6	14.5
	7.6	9.0	10.3	73.1	100

[1]Numerator – number of plants, thousands per ha
[2]Denominator – percentage

TABLE 3. Number of young plants of *Artemisia lessingiana* and their survival in desert conditions. Number of plants per square meter (Ulguli village, yearly averages)

Grazing	Number				Died %	
	Beginning of 1982	End of 1982	1983	1984	In summer	In winter
Light	128	25	9	9	80.5	36.0
Intensive	76	15	7	6	82.9	46.2
Very intensive	31	12	6	4	81.5	46.7

TABLE 4. Root mass formation in relation to level of pasture use, tons per hectare 2006

Level of grazing	Depth in soil layers, cm		
	0–10	10–20	20–30
Weak	4.01	0.65	0.46
Intensive	2.34	0.56	0.34
Very intensive	0.74	0.48	0.29

weight of *Artemisia* shrubs in dry year and on intensively grazed pastures was not more than 7–8 grams, while the average weight of the same plant at light grazing levels was 13 grams, increasing to 20 grams in wet years.

A variety of nutritive elements – carbohydrates, starch, mono- de-sugar and hemicelluloses – play a part in the development of pasture plants. In nourishing the sprouting process, the most important parts of the plant are the root system and the lower plant which do not freeze in winter and

TABLE 5. *Artemisia terrae-albae* average shrub weights (grams per plant of above ground dry matter) at different levels of grazing intensity in wet and dry years (Aydarly)

Level of grazing	Dry year – 2006 Repetitions of weighing 1	2	3	4	5	6	Average weight in grams	Wet year – 2004 Repetitions of weighing 1	2	3	4	5	6	Average weight in grams
Medium	12	14	13	15	9	13	13	20	19	18	20	22	21	20
Intensive	7	10	8	9	8	8	8	12	14	14	11	12	10	12
Very intensive	6	9	5	6	7	7	7	8	10	10	11	9	12	10

TABLE 6. Impact of grazing intensity on the composition of digestible sugar and starch in plants, shown in % (*Artemisia-Anabasis salsa* pastures, winter 1999)

	Artemisia		*Anabasis salsa*	
	Root	Root stem	Root	Root stem
Use	Percentage digestible sugar			
Weak	4.03	4.01	5.74	5.17
Intensive	3.64	3.21	5.13	4.72
Very intensive	3.09	2.77	4.40	4.11
	Percentage digestible starch			
Weak	0.32	0.77	0.46	0.99
Intensive	0.24	0.64	0.38	0.79
Very intensive	0.20	0.47	0.32	0.61

provide additional nutrition. In order to define the condition of plants at the study sites we conducted a study of the composition of plant nutritional reserves, summarized in Table 6.

Table 6 shows that in the species studied here the largest portion of digestible sugar was concentrated in the root and a slightly smaller amount was located in the upper part of the root system. A decrease of reserve carbohydrates of up to 30% could be found on pastures with extensive grazing and consequently these plants have very weak resistance to cold weather as well as delayed dates for sprouting in spring or for reaching peak biomass later in the growing season. Accumulation of starch in *Artemisia* and *Anabasis salsa* is concentrated in the stems of the root. Intensive grazing on these plants decreases the starch in the plant by 19–28% which could affect its longevity and productivity.

3. Impact of pasture grazing intensity upon the species composition of vegetation

The results of the research work conducted between 1999–2004 demonstrate that two different processes are taking place on Kazakhstan's pastures. Pastures around villages are under intensive grazing pressure and therefore most of them are quite degraded, while pastures distant from settlements have almost completely regenerated. However almost all types of regenerated pastures are now covered by moss which has quite a negative impact on the vegetation. The issue of changes in the vegetation communities of pastures has been studied by us on several different types of pastures: *Artemisia*-ephemeron, *Artemisia*-grass-ephemeron, *Stipa-Artemisia* and also on shrub-*Agropyrum*- ephemeron pastures with saksaul (*Haloxylon persicum*). Monitoring of dynamics of number of the dominant species was carried out depending on pasture use regime over the last 5 years (2000–2004). Some of the main results are presented in Table 7.

Table 7 shows that after five years of intensive use the quantity of dominant species (*Artemisia terrae-albae*) decreased while the quantity of ephemerons *(Eremopyrum orientalis)* increased by 70% in desert pastures. In contrast, under a light grazing regime there is some increase of the wormwood (*Artemisia terrae-albae*) by 15%, and the quantity of ephemerons *(Eremopyrum orientalis)* is stable.

Changes that have taken place on *Artemisia*-grass-ephemeron semi-desert pastures are reflected in Table 8. Foothill and semi-desert zones which are characterized by dominant subshrubs such as *Artemisia lessingiana*, and grasses such as *Stipa capillata*, exhibit the same trends in changing species

TABLE 7. Impact of grazing pressure on *Artemisia*-ephemeron desert pastures, measured in terms of the number of dominant plants per square meter

	Level of grazing intensity		
Years	Light	Intensive	Very high
2000	4.0[1]/76.6[2]	2.7/106.3	1.5/117.3
2001	4.6/88.0	2.8/123/3	1.5/164.6
2002	4.6/69.4	2.7/105.6	1.1/197.7
2003	5.3/78.8	2.9/129.3	1.1/179.6
2000	5.2/96.4	2.9/14438	0.7/248.3

[1]Numerator – *Artemisia terrae-albae* plants per square meter
[2]Denominator – *Eremopyrum orientalis* plants per square meter

TABLE 8. The impact of different levels of grazing on the number dominant plants per square meter in semi-desert pastures

Years	Level of grazing intensity		
	Light	Intensive	Very high
2000	5.3[1]/3.8[2]	3.6/2.4	1.8/2.0
2001	5.2/4.2	3.9/2.6	1.8/2.0
2002	5.8/3.8	3.2/2.0	1.1/1.6
2003	6.1/4.1	3.2/2.0	1.3/1.6
2000	5.8/4.2	3.0/2.0	1.0/1.3

[1]Numerator – *Artemisia lessingiana* plants per square meter
[2]Denominator – *Stipa Capillata* plants per square meter

composition as desert zones. Intensive grazing decreases the number of even grazing resistant plants like *Stipa* and *Artemisia lessingiana* declines to about half of its prevalence at pasture sites which have been less intensively grazed.

At the same time we can see an increase in *Carpoceras glabrata, Artemisia scoparia* and other less palatable plants (Table 9). Table 9 shows that intensively grazed sites are characterized by increasing quantity of *Peganum harmala* (*Artemisia* ephemeron type of pastures) and *Artemisia scoparia* (in *Artemisia-stipa*-ephemeron type of pastures).

In order to study the impact of grazing on the quantity of dominant plants in different plant communities, we have been studying and monitoring pasture sites in Moinkum Desert over the last five years on pastures characterized mainly by shrubs-Agropyrum-ephemeron with saksaul (*Haloxylon persicum*). The results of this work can be found in Table 10.

Table 10 demonstrates the different way that the level of grazing intensity affects different plant species. Intensive grazing immediately affects the quantity of *Kohia*, which totally disappears from the grass stand if grazing is quite intensive. *Agropyrum* is more resistant but still declines with persistent high levels of grazing pressure.

The previous data (summarized in Tables 7–10) demonstrates that unsystematic, intensive grazing changes the species composition of vegetation communities. For example, in the first your of monitoring, *Artemisia*-ephemeron pasture could be classified as *Carpoceras glabrata*-ephemron-*Artemisia* type of vegetation community, but five years later these same pastures had been transformed into vegetation communities dominated by *Peganum harmila*, an unpalatable plant, with a much diminished presence of other species. Table 11 summarizes the typical direction of change in species composition in response to very intensive grazing of the plant communities examined in

TABLE 9. Grazing impacts upon the quantity of *Peganum* and *Artemisia scoparia*, thousand species per h/a

	Types of pasture					
	Artemisia ephemeron desert pastures			*Artemisia-stipa*-ephemeron dry steppe pastures		
	Peganum harmala			*Artemisia scoparia*		
Years	Weak	Intensive	Very intensiv	Weak	Intensive	Very intensive
2000	2.3	3.6	12.8	3.6	5.8	16.6
2001	2.6	3.9	14.6	2.9	5.6	19.3
2002	1.8	4.2	21.3	3.4	7.9	17.8
2003	0.7	3.9	27.7	2.7	7.6	24.3
2004	1.0	4.6	46.8	1.3	9.1	38.6

TABLE 10. Grazing impact upon quantity of dominant species in shrubs-*Agropyrum*-ephemeron with saksaul in the Moinkum Desert, number of plants per square meter

	Level of grazing intensity		
Years	Light	Intensive	Very high
2000	2.6[1]/4.4[2]	1.8/2.3	1.6/1.3
2001	3.1/3.8	2.1/2.4	0.7/1.0
2002	3.6/4.1	1.6/2.1	–/1.4
2003	4.5/4.4	1.9/2.0	–/1.0
2004	5.3/4.7	1.3/1.6	–/1.0

[1]Numerator – number of *Kohia prostrata* plants
[2]Denominator – number of *Agropyrum*

TABLE 11. Modification of studied pasture types as a result of grazing

Monitoring sites	Use intensity	Pasture modification
Desert zone, Aydarly	Very intensive	*Peganum*
	Intensive	*Carpoceras glabrata -Artemisia-Peganum*
	Weak	*Artemisia*-ephemeron
Semi-desert zone, Ulguli	Very intensive	*Artemisia- Carpoceras glabrata*
	Intensive	*Artemisia*-grasses- *Carpoceras glabrata*
	Weak	*Stipa-Artemisia- Festuca*
Dry steppe zone, Shien	Very intensive	*Artemisia*
	Intensive	*Artemisia-Festuca-Stipa*
	Weak	*Stipa –Festuca- Artemisia*
Moinkum, sand pastures	Very intensive	Shrubs- *Carpoceras glabrata*
	Intensive	Shrubs-*Agropyrum-Carpoceras glabrata*
	Weak	*Artemisia-Kohia-Agropyrum*

this study. Table 11 shows that the species composition of all pasture types is modified by high levels of unsystematic use.

4. Impact of intensive pasture use on the volume of forage production

It is important to consider the total volume of forage output in defining an optimal system for pasture use. It must be remembered that weather conditions influence pasture productivity as well as human impact.

The impact of pasture use intensity upon the dynamics of fodder mass accumulation was studied over five years at six monitored sites with different pasture types along two north-south transects. The sites along each transect are listed below from south to north:

Alatau Mountains-Balkash basin (Almaty Oblast) transect (Plate 5):

- *Stipa-Artemisia* (foothills, dry steppe).
- *Artemisia*-grass mixture (semi-desert),
- *Artemisia*-ephemeron (desert),

Moinkum desert-Betpakdalla plains (Jambul Oblast) transect (Plate 8):

- shrubs-*Carpoceras glabrata*-ephemeron with saksaul (Moinkum sands),
- *Anabasis salsa-Artemisia* (right bank of the Chu River)
- *Artemisia*-grass pastures in Betpakdalla desert (Maitokken valley).

***Stipa-Artemisia* (foothills, dry steppe)** was monitored in the vicinity of Shien village. Pastures within a one km radius of the village had been heavily grazed. Average palatable dry biomass over a five year period was: spring: 2.0 c/ha, summer 3.1 c/ha, autumn 2.3 c/ha and in winter 1.8 c/ha. The plant biomass of these pastures increased at a 2 km radius from the village. Average palatable dry biomass for the last 5 years comprised: spring: 2.5 c/ha, summer 3.8 c/ha, autumn 2.8 c/ha and in winter 2.1 c/ha. By 2004 pastures located far from the village were regenerating and their dry biomass was: spring: 2.8 c/ha, summer 5.2 c/ha, autumn 3.7 c/ha and in winter 2.6 c/ha.

***Artemisia*-grass mixture (semi-desert)** was monitored around Ulguly village (Plate 6). *Artemisia*-grass type of pastures at one km from the village had been degraded and modified. Their palatable biomass at different seasons was: spring-1.5 c/ha, summer-2.7 c/ha, autumn-1.8 c/ha, winter-1.1 c/ha. The situation improved at 2 km radius from the village. Average palatable dry biomass at 2 km from the village for the last 5 years was: spring: 1.7 c/ha, summer 4.1 c/ha, autumn 2.8 c/ha and in winter 2.1 c/ha. In 5 km distance the pastures were in much better condition: spring-2.9 c/ha, summer-5.2 c/ha, autumn-4.3 c/ha, winter-3.3 c/ha. Relatively high pasture yield in this zone is explained by the increase of *Artemisia lessingiana* sub-shrub which replaces *Stipa capillata*

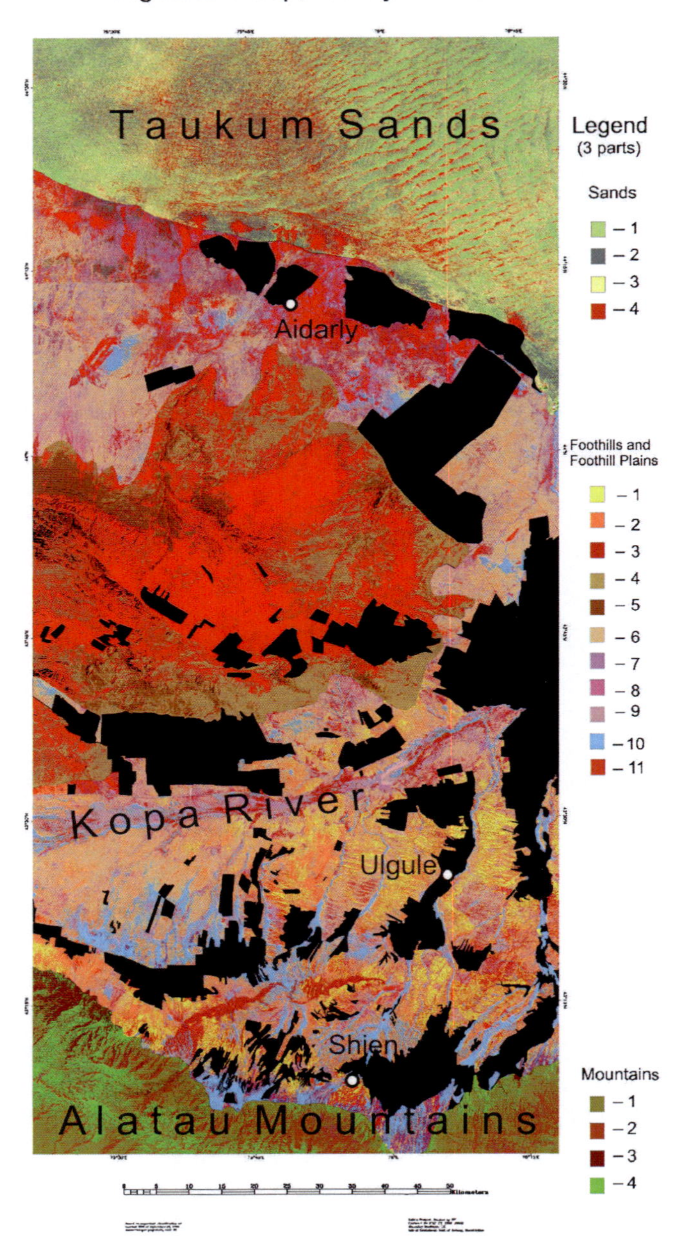

Plate 5: Vegetation Map, Almaty Transect. Map compiled by Sayat Temirbekov

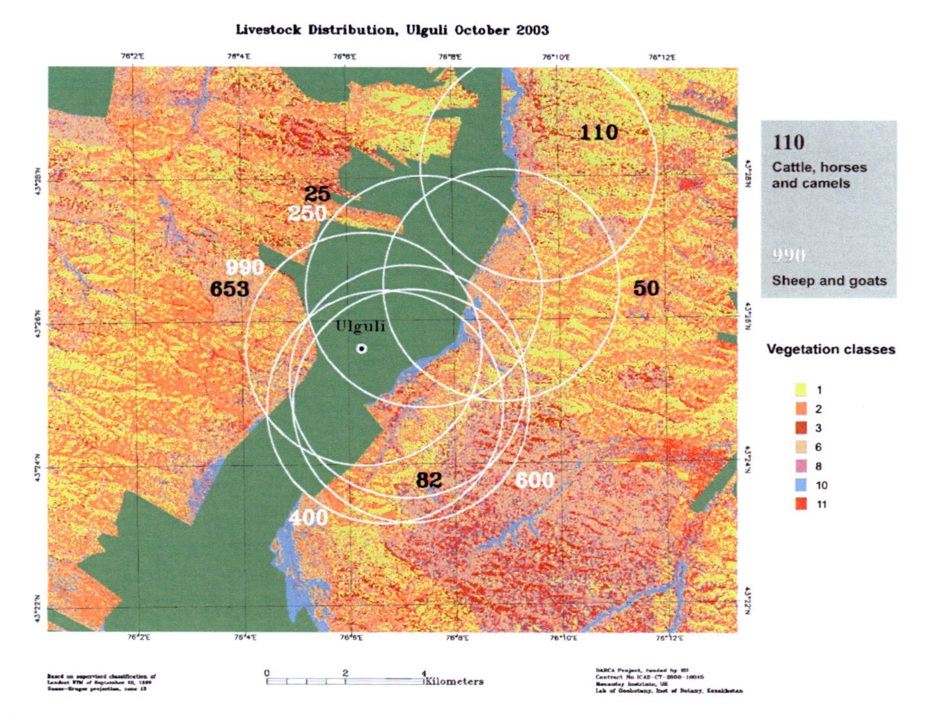

Plate 6: Livestock Distribution, Ulguli October 2003. Vegetation map compiled by Sayat Temirbekov; flock distributions by Roy Behnke and Aidos Smailov

under heavy grazing. However, this transition can be accelerated with negative consequences by continuous unsystematic grazing leading to the eventual disappearance of *Artemisia lessingiana,* which is a valuable fodder species.

***Artemisia*-ephemeron (desert)** pastures (exemplified by conditions around Aydarly village, Plate 7) were heavily exposed to unsystematic grazing. At 1 km distance from the village the grass stand had been totally changed and consisted mainly of annual species: spring-0.6 c/ha, summer-1.1 c/ha, autumn-0.2 c/ha, winter-0.2 c/ha. Due to the presence of sub-shrubs, seasonal yield looks different in 2 km distance from the village: spring-0.8 c/ha, summer-2.0 c/ha, autumn-0.9 c/ha, winter-0.7 c/ha. Finally, at 5 km the seasonal values are: spring-1.8 c/ha, summer-4.6 c/ha, autumn-4.9 c/ha, winter-2.6 c/ha on fully regenerated *Artemisia*-ephemeron pastures.

Outside Aydarly the mixed shrub-grass pastures of the Sarytaukum sands traditionally served as winter pastures. Seasonal dynamics in this area was: spring-3.0 c/ha, summer-4.5 c/ha, autumn-4.6 c/ha, winter-3.0 c/ha. These dune pastures were important because valley and foot hill pastures were on average under snow for 45 days each year, during which time grazing remained available in the sand dune pastures.

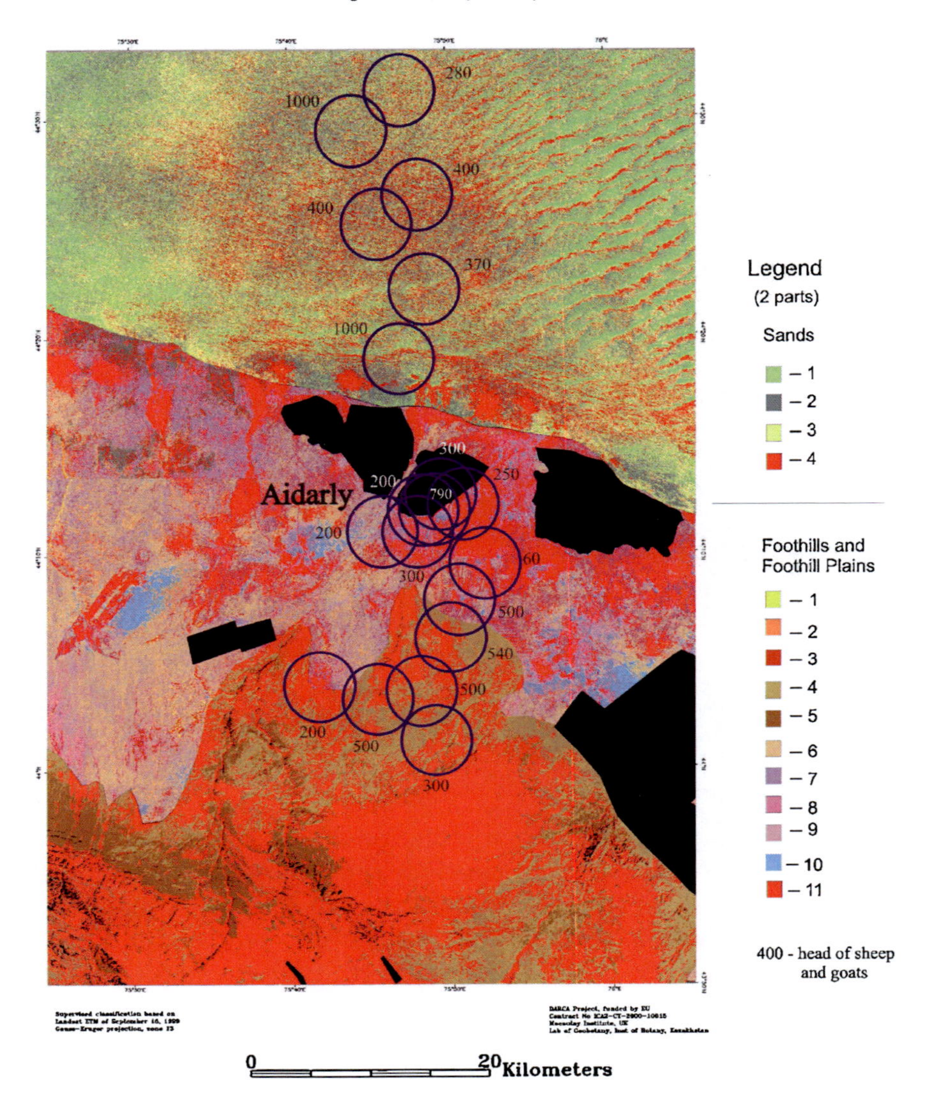

Plate 7: Flock Distributions and Vegetation, Aidarly January 2003. Vegetation map compiled by Sayat Temirbekov; flock distributions by Roy Behnke and Aidos Smailov

We now examine seasonal biomass production along the second transect, in the Moinkum desert-Betpakdalla plains (Jambul Oblast), starting with vegetation types at the southern end of the transect (Plate 8 and 9).

Shrubs-*Carpoceras glabrata*-ephemeron pastures with saksaul *(Haloxylon persicum)* located in the Moinkum sands were mainly used in spring-summer-autumn times without any rotation of grazing sites. As a result of

BOX 1. Legend for the classification of vegetation on the Almaty transect, Plates 5, 6 and 7

Vegetation of sands (northern part of the study area) Plates 5 and 7

1. Agropyron-bush (*Calligonum sp., Agropyron fragile*). and izen-sagebrush (*Artemisia terrae- albae, Kochia prostrata*) sands
2. Eurotia with sagebrush and ephemeroids sands (*Ceratoides papposa, Artemisia terrae-albae,Carex physodes, Carex pahistilis, Anisantha tectorum*) sands
3. Sagebrush with eurotia, izen, ephemeroids (*Artemisia terrae-albae, Ceratoides papposa, Carex physodes, Carex pahistilis, Anisantha tectorum*) sands
4. Overgrazed areas of low vegetation coverage and bare sands or dominated by ruderal sagebrush vegetation (*Artemisia scoparia, A. songarica*)

Vegetation of foothills and foothill plains (central part of the study area) Plates 5, 6 and 7
Foothill desert steppes
Feather grass pastures on upper and dissected foothills

1. Forb-feather grass pastures on uppermost foothills (*Stipa capillata, Ziziphora bungeana, Phlomis goraninovii*); sagebrush-feather grass pastures on upper undulating foothills (*Stipa capillata, Artemisia sublessingiana, Festuca valesiaca*); feather grass pastures on northern slopes of dissected loess foothills (*Stipa capillata*)

Ephemeroid-grass-sagebrush pastures on undulating foothills and foothill plains

2. Ephemeroid-grass-sagebrush pastures and sagebrush pastures with grasses (*Artemisia sublessingiana, Stipa capillata, Poa bulbosa, Carex pahistilis*)
3. Ephemeroid-grass-sagebrush pastures and sagebrush pastures with grasses on stony soils (*Artemisia sublessingiana, Stipa capillata, Festuca valesiaca, Carex pahistilis, Koeleria cristata*)

Foothill deserts
Ephemeroid-sagebrush pastures with bushes in hilly upland

4. Ephemeroid-sagebrush pastures with bushes
5. Bush-ephemeroid-sagebrush pastures

Sagebrush pastures with grasses and ephemeroids on gently sloping plains

6. Sagebrush pastures with grasses and ephemeroids (*Artemisia sublessingiana, A. terrae-alba, Stipa capillata, Poa bulbosa, Carex pahistilis*)
7. Sagebrush-kereuk (*Artemisia terrae-alba, A.turanica, Salsola rientalis,Ceratocarpus arenarius*) and kereuk-sagebrush (*Salsola orientalis, Artemisia terrae-alba, Ceratocarpus arenarius*) pastures

Halophytic sagebrush and perennial saltwort pastures of low riverine plains

8. Complexes of *Camphorosma monspeliaca-Artemisia schrenkiana* and *Climacoptera brachiata* communities
9. Perennial saltwort pastures (*Halocnemum strobilaceum*)

Meadow pastures and pastures with additional water supply along water ways, river valleys

10. Meadow and bush vegetation along water ways, river valleys (*Tamarix sp., Phragmites australis, Lasiogrostis splendens*); reed thickets (*Phragmites australisi*)

Degraded vegetation

11. Trampled down bare surfaces and thickets of *Atriplex tatarica*; Sagebrush-ebelek (*Artemisia sublessingiana, Ceratocarpus utriculosus, C. arenarius*) and ebelek (*Ceratocarpus utriculosus, C. arenarius*) vegetation; overgrazed areas with *Climacoptera brachiata* and *Ceratocarpus sp.*; overgrazed areas with *Peganum harmala*; abandoned croplands with *Polygonum aviculare, Bromus tectorum*

Vegetation of Mountains (southern part of the study area) Plate 5

1. Forb-grass and grass-forb steppes *(Stipa capillata, Festuca valesiaca, Koeleria gracilis, poa angustifolia, Ziziphora bungeana,Origanum vulgare, Phlomis goraninovii, Artemisia dracunculus)*
2. Forb-grass steppes on stony soils
3. Petrophytic forb steppes on stony soils and bushes on rock outcrops
4. Bush thickets

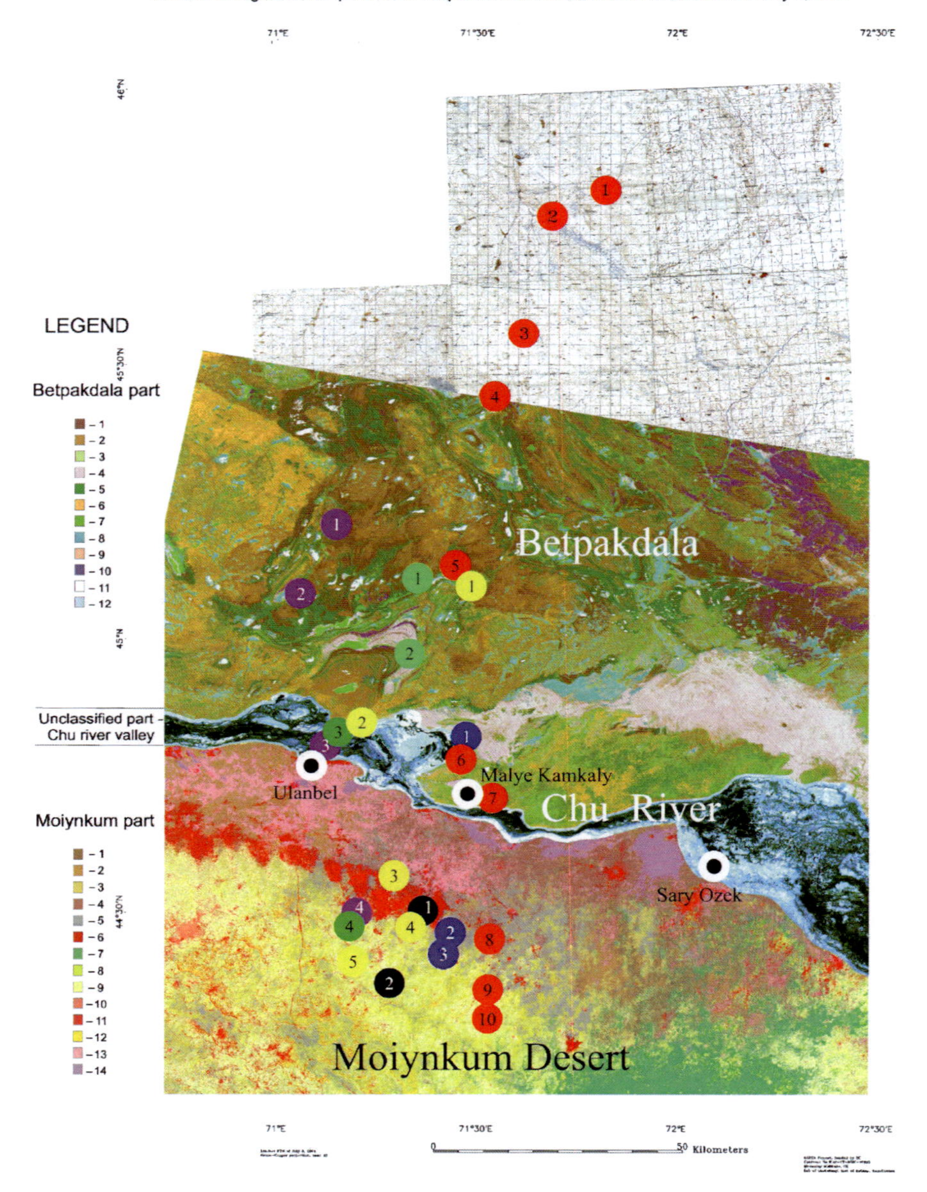

Plate 8: Flock Migrations in the Moiynkum Area, 2002–2003. Vegetation map compiled by Sayat Temirbekov; flock distributions by Roy Behnke and Aidos Smailov

unsystematic use of pastures around the settlement in the last year of our study, the grass stand underwent great changes and nearly lost the capacity to regenerate. The following data was received for palatable yield in spring:

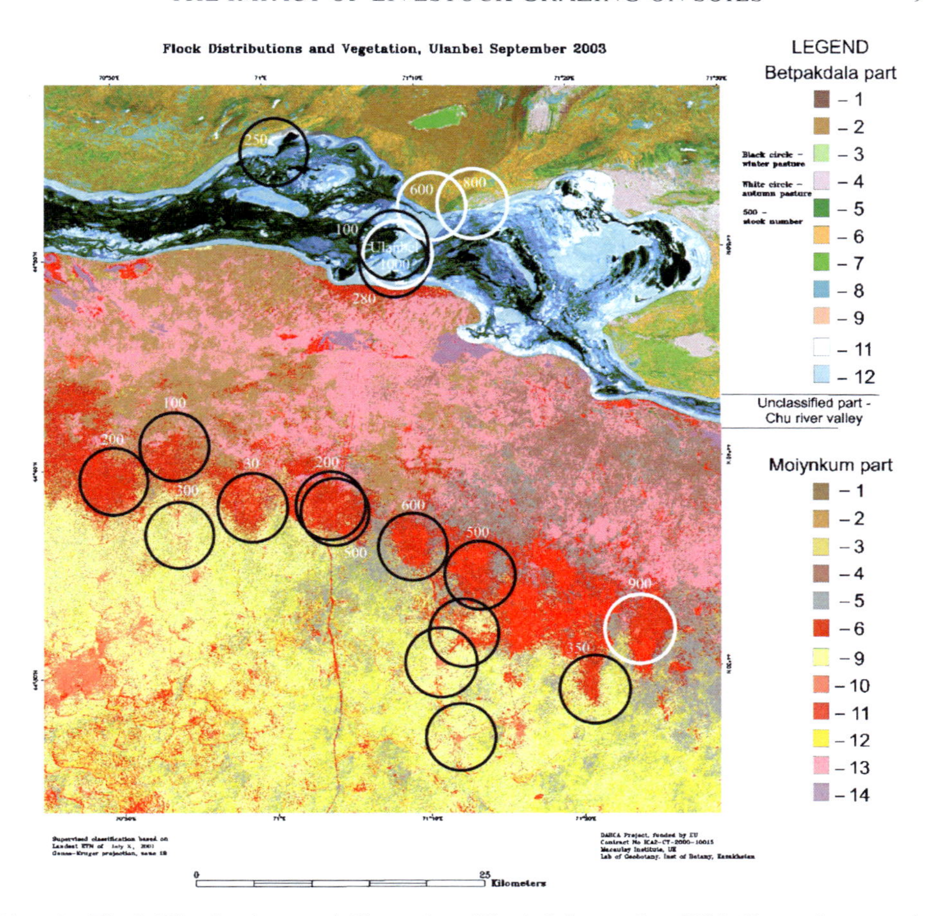

Plate 9: Flock Distributions and Vegetation, Ulanbel September 2003. Vegetation map by Sayat Temirbekov; flock distributions by Roy Behnke and Aidos Smailov

spring-2.2 c/ha, summer-5.7 c/ha, autumn-3.6 c/ha, winter-2.7 c/ha. Palatable yield was mainly formed from annual plants and *Agropyrum* since there were very few sub-shrubs (*Artemisia* and *Kohia*, 1–3 plants per m^2) and their proportion in the total yield was not more than 23–29%. In 2.5 km distance from the barn pasture condition gradually improved because of the greater prevalence of sub-shrubs. However the soil in this area was quite degraded and evidence of intensive use by livestock was obvious. This area had the following indicators of palatable dry mass: spring-3.3 c/ha, summer-6.9 c/ha, autumn-5.3 c/ha, winter-4.4 c/ha.

There was quite a different picture in 5 km distance from the main livestock concentrations, where pastures were regenerating. The number of sub-shrubs increased to 7–9 in one m^2 in comparison to 1–3 sub shrubs in degraded pastures. There was also significant change in accumulated biomass

BOX 2. Legend for the classification of Betpakdala vegetation in Plates 8 and 9

Low mountain vegetation on stony soils with rock outcrops dominated by black salsola, sagebrush-black salsola and petrophytic communities
Vegetation coverage 30–50%

1. *Artemisia turanica*-black salsola (*Salsola arbusculaeformis, Artemisia turanica, Artemisia semiarida, Catabrosella humile*) and petrophytic communities (*Ephedra diatachya, Artemisia juncea, A. sublessingiana*).

Vegetation on plains on loamy-cobble soils dominated by sagebrush-black salsola
Vegetation coverage 30–50%

2. Complex of ephemeral-sagebrush-black salsola (*Salsola arbusculaeformis, Artemisia turanica, Artemisia terrae albae, Eremopyrum orientale, Diptychocarpus strictus, Astragalus fiicaulis*) with *Nanophyton erinaceum-Anabasis salsa*.

Vegetation on plains on loamy sand soils dominated by *Salsola orientalis*-sagebrush communities
Vegetation coverage 30–50%

3. *Salsola orientalis*-sagebrush (*Artemisia terrae-albae, Salsola orientalis*).

Vegetation coverage 50–70%

4. *Salsola orientalis*-sagebrush (*Artemisia terrae-albae, Salsola orientalis, Stipa szowtsiana*) communities with saxaul (*Haloxylon aphyllum*).

Vegetation on loamy soil dominated by *Artemisia turanica* and ephemerals
Vegetation coverage 50–70%

5. Ephemeral-*Artemisia turanica* (*Artemisia turanica, Artemisia terrae-albae, Eremopyrum orientale, Astragalus filicaulis, Psammogeton setifolium*).

Vegetation coverage 30–50%

6. Ephemeral (*Ceratocarpus utriculosus, Tetracme quadricornis, Alyssum turkestanicum, Eremopyrum orientale*) communities with scuttered shrubs and semishrubs (*Salsola orientalis, Artemisia terrae-albae, Artemisia turanica*).

Vegetation in depression with eolian sand dominated by saxaul woodlands
Vegetation coverage 60–80%

7. Saxaul woodland with undergrowth of eurotia-*Salsola orientalis*-sagebrush (*Artemisia semiarida, A. turanica, Salsola orientalis, Cereatoides papposa*) and ephemeral (*Petrosimonia sibirica, Ceratocephalus falcatus, Eremopyrum orientale*) communities.

Vegetation on saline soils dominated by halophytic communities
Vegetation coverage 30–50%

8. Anabasis salsa (Anabasis salsa, Nanophyton erinaceum) communities.
9. Anabasis salsa-sagebrush (Artemisia terrae-albae, Anabasis salsa).
10. Complex of *Atriplex cana*, *Artemisia schrenkiana-Atriplex cana* communities.

Vegetation coverage - below 10%

11. Sparse saltwort communities on takyrs.
12. Sparse halophytic communities on solonchaks.

BOX 3. Legend for the classification of Moiynkum vegetation in Plates 8 and 9

Vegetation of plains with takyr-like soils with eolian sand dominated by saxaul woodlands

Vegetation coverage 50–70%

1. Saxaul woodland (*Haloxylon aphyllum, Salsola orientalis, Artemisia terrae-albae, A. turanica, Eremopyrum triticeum*).

Vegetation coverage 30–50%

2. Sparse saxaul woodland with ephemeral-sagebrush-Salsola orientalis (*Haloxylon aphyllum, Salsola orientalis, Artemisia terrae-albae, A. turanica, Eremopyrum triticeum*) undergrowth
3. Saxaul woodland with wheat grass-eurotia-sagebrush undergrowth (*Haloxylon aphyllum, Ceratoides papposa, Artemisia terrae-albae, Agropyron fragile, Carex physodes*) on thick eolian sand.
4. Salsola orientalis-saxaul, annual saltwort-saxaul (*Haloxylon aphyllum, Salsola orientalis, Clima-coptera brachiata, Petrosimonia sibirica*) communities.

Vegetation of plains with thin eolian sand dominated by eurotia communities

Vegetation coverage 30–50%

5. Sagebrush-eurotia and eurotia-sagebrush (*Ceratoides papposa, Artemisia terrae-albae, Kochia prostrata, Alyssum desertorum, Astragalus filicaulis*) communities.

Vegetation coverage 10–20%

6. Weed-eurotia-sagebrush (*Artemisia terrae-albae, Ceratoides papposa, Cousinia sp., Bromus tectorum*) communities. Overgrazed vegetation.

Vegetation of hummocky, hummocky-ridge and honeycomb sands dominated by saxaul, psammophytic shrub and sagebrush communities

Vegetation coverage 50–70%

7. Ephemeral-summer cypress-sagebrush (*Artemisia terrae-albae, Artemisia leucoides, Kochia prostrata, Calligonum caputmedusae, Hyalea pulchella, Bromus tectorum*), ephemeral-wheat grass-sagebrush-shrub (*Calligonum caputmedusae, Haloxylon aphyllum, Astragalus brachypus, Ceratoides papposa, Artemisia terrae-albae, Agropyron fragile, Hyalea pulchella, Bromus tectorum*) communities.
8. Saxaul (*Haloxylon aphyllum, H. persicum, Carex physodes*) communities in wide dune valleys.

Vegetation coverage 30–50%

9. Sparse psammophytic shrub-saxaul (*Haloxylon persicum, H. aphyllum, Calligonum caput-medusae, Ammodendron argenteum, A. terrae-albae, Agropyron fragile, Stipa szowitsiana*) and psammophytic shrub-sagebrush with saxaul (*Artemisia terrae-albae, A.songarica, Calligonum caputmedusae, Calligonum leucode, Haloxylon aphyllum, Astragalus brachy-pus, Ceratoides papposa*) communities on hummocky-ridge sands.
10. Saxaul woodland with sagebrush-ephemeral undergrowth (*Haloxylon persicum, Bromus tectorum, Artemisia terrae albae, Kochia prostrata*). Overgrazed vegetation.
11. Saxaul woodland with weed-ephemeral undergrowth (*Bromus tectorum, Artemisia leucoides, Londesia eriantha*). Overgrazed vegetation.
12. Sagebrush-psammophytic shrub with saxaul (*Calligonum caputmedusae, Caligonum leucocladum, Astragalus brachypus, Artemisia terrae-albae, A. songarica, Artemisia leucodes, Agropyron fragile*) communities on smoothed hummocky sands.

Halophytic vegetation on takyr-like clayey soils and solonchaks

Vegetation coverage 10–20%

13. *Anabasis salsa* communities.

Vegetation coverage 10%

14. Sparse halophytic communities on solonchaks.

BOX 4. Notes for Plates 5–9

Plate 5, ***Vegetation Map Almaty Transect,*** shows the location of three study villages: Aidarly in the north on the fringe of the Taukum sands, Ulgule in the central plains, and Shien in the foothills of the Alatau Mountains.

Plate 6, ***Livestock Distribution, Ulguli October 2003,*** shows the position of flocks and herds around Ulgule village. The white circles represent the three kilometer daily grazing radius of a typical flock or herd; for visual simplicity, several flocks at the same location are represented by a single circle. The stock numbers represented by each circle are indicated numerically on the map, white numbering representing small ruminants and black numbering giving the total head of cattle, horses and camels in an area. The overlapping circles and large flock sizes on this map suggest considerable grazing pressure around Ulgule village. Over 2200 head of sheep and goats and about 800 large stock resided permanently within 5 km of the center of the village in 2003. About 42% of all grazing by village-owned sheep and goats occurred within a five kilometer radius of the village, at a seasonally constant stocking rate of about 35 hectare per sheep.

Plate 7, ***Flock Distributions and Vegetation, Aidarly 2003,*** illustrates the size and position of flocks around Aidarly village in the winter of 2003.For each flock, black circles indicate a typical daily grazing radius;

numbers give flock size in total head of sheep and goats. In 2003 about 40% of all grazing by village-owned sheep and goats occurred within a five kilometer radius of the village; stocking rates for small ruminants were lowest around the village in winter (34 ha/sheep unit) and averaged 21 ha/sheep unit in the other seasons.

Plate 8, ***Flock Migrations in the Moiynkum Area, 2002–2003,*** shows the location of three study villages along the Chu River – Ulanbel, Malye Kamkaly and Sary Ozek – and tracks the movements of all migratory flocks along this section of the Chu. Each circle on Map 4 represents a flock's three kilometer grazing radius, in this instance around a temporary camp or seasonally inhabited building. Each series of colored circles – red, blue, yellow, etc. – depicts the movements/camping locations of a single flock over the course of one year. The numbers on the colored circles indicate the sequence of camp movements. The basic direction of flock movement is from north to south; no flock moves contrary to this pattern. But the amplitude of movement is highly variable, as is the number of moves in a yearly cycle. Some flocks – the flock represented by black circles 1 and 2, for instance – move only twice and over a short distance, shifting from summer pastures to winter quarters and back again, all within the same vegetation zone. At the other extreme, the most mobile flock in the area makes 10 major moves, covering a round trip distance of about 350 km. In addition to resting their home pastures, these mobile flocks take advantage of seasonal differences in temperature, diminished numbers of insect pests in some areas, and variable forage quality and quantity across three major ecological zones. These three zones are the southern sand deserts (winter), the Chu river valley (spring when heading north and again in autumn upon returning south) and the clay plains of the Betpakdala (summer).

Plate 9, ***Flock Distributions and Vegetation, Ulanbel September 2003,*** illustrates the location in September 2003 of flocks owned by residents of Ulanbel, the most westerly and largest Chu River community in this study. In Ulanbel, very few animals stay in the village year round – under 400 head out of a total of about 10,000 small ruminants owned by villagers. The majority of village-owned flocks live permanently in the vicinity of stock enclosures and at wells (indicated by black circles with a 3 km radius) 20–40 km south of the village in the sand dunes of the Moinkum desert. Depicted as white circles on Map 3 are the locations of migratory flocks at the time of the livestock census in September of 2003.

quantity in different seasons: spring-3.3 c/ha, summer-6.9 c/ha, autumn-5.3 c/ha,winter-4.4 c/ha of dry palatable phytomass.

The monitored site for ***Artemisia*-grass pastures in Betpakdalla desert (Maitokken valley)** is located 220 km north of the site in the Moinkum sands. These pastures are remote from any permanent settlement and are used by individual shepherds in summer time. After ten years of little or no grazing, the resource potential of these Artemisia-grass pastures has fully regenerated both in terms of species composition and fodder yield. The dynamics of fodder mass accumulation follows a classic pattern. Pasture yield is formed from ephemerons, *Carpoceras glabrata* and sub-shrubs in spring and in the first half of the summer, and in the second half of summer and in autumn the yield is formed from sub-shrubs. Maximum yield on these type of pastures is reached by the mid of June (4.7 c/ha), and in autumn due to disappearance of annual plants the yield decreases by 30–35%. It should be mentioned that the last 5 years of pasture use did not negatively affect pasture productivity but that sites at 2.5 km from water source were already undergoing changes that indicated a trend towards degradation.

Table 12 summarizes the changes in the species composition of vegetation communities caused by grazing at monitoring sites located along the Alatau-Balkhsh.

Table 12 shows that all climatic zones experience an increase of unpalatable plants if the grazing is intensive. In the dry steppe and semi-desert degraded pastures include *Carthamnus* and *Artemisia scoparia*; desert

TABLE 12. The proportion of palatable pasture biomass at different levels of grazing intensity in the summer of 2003

Zones	Distance from livestock concentration points	Palatable perennials - %	Palatable annuals - %	Not completely palatable perennials and annuals - %
Foothill dry steppe (Shien)	1.0	46.2	31.7	22.1
	2.5	59.6	37.6	2.8
	5.0	67.7	30.4	1.9
Semi desert (Ulguly)	1.0	66.2	16.2	17.6
	2.5	65.6	30.6	3.8
	5.0	67.2	27.3	5.5
Desert (Aydarly)	1.0	15.4	30.8	53.8
	2.5	32.6	41.6	25.8
	5.0	62.3	31.3	6.4
Sand desert (Sarytaukum)	Regenerated	80.6	18.4	1.0

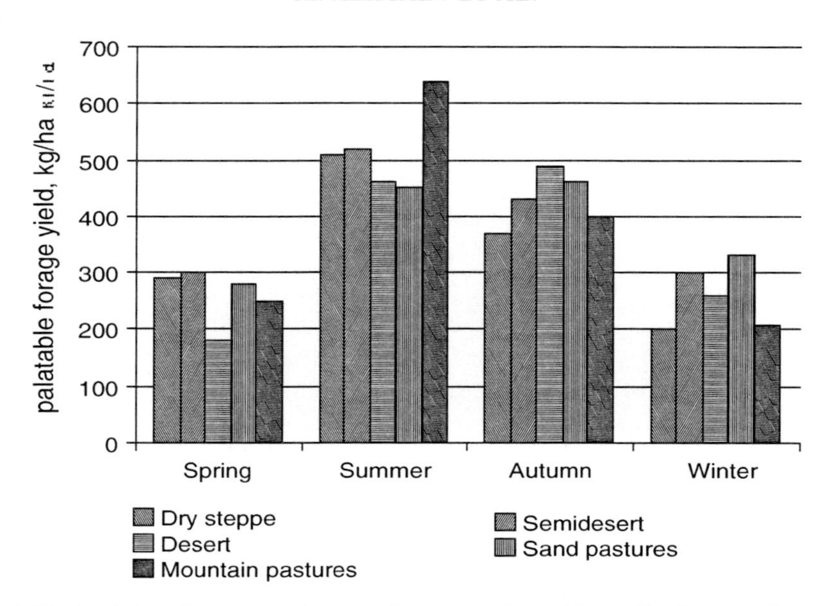

Figure 1. Productivity of regenerated seasonal pastures along Alatau- Taukum sands transect

pastures have an increase in *Peganum* communities among dominating vegetation species. The regenerated sand pastures have about 1.0% of unpalatable *Astragalus* among the grass stand. Palatable perennials include *Stipa capillata, Festuca species* and *A. terrae-albae* in the foothills and dry steppe, *Stipa capillata* and *A. sublessingiana* in the semi desert, *A. terrae-albae, Kochia prostrate* and *Ceratoides species* in the desert with the addition of *Carex physodes* in the sand desert. Palatable annuals include species of *Bromus* and *Triganellam* in the foothills, *Papver* and *Tulipa* in the semi desert and *Eremopyrum* and *Ceratocarpus* in the desert and sand desert.

The seasonal dynamics of pasture biomass accumulation helps to explain the necessity of moving livestock throughout the year. Maximum pasture biomass accumulation occurs in the mountains, foothills-dry steppe and semi-deserts in summer, at the edge of the sand deserts in autumn, and within the sand deserts in winter. These seasonal fluctuations in peak pasture production support the traditional principle of using a combination of different pasture types as a single production system based upon the systematic exploitation of spatial and seasonal variations in fodder productivity (Figure 1).

5. Chemical composition of forage in relation to pasture use regime

Pastures in the study area are generally rich in nutritious elements. For example, the raw protein composition of *Artemisia*-ephemeron pastures in spring and summer could easily be compared to lucerne hay. However, the chemical composition of fodder from pastures is not stable and changes depending on vegetation species composition, season, rainfall and weather conditions in a given year, plant condition and many other factors.

We attempted to record the nutritional value of fodder plants of pastures located along Alatau-Balkhsh (Plate 5) and Moinkum-Betpakdalla (Plate 8) transects in different seasons of use and depending on pasture condition.

Artemisia pastures are the most valued as winter grazing for animals. In winter the palatability of *Artemisia* is higher than all other species, even more than *Kohia*. The nutritional value of the plant changes as the plant grows and according to meteorological conditions, declining in dry years as the amount of protein, carbohydrates and oil decreases while there is increase in cellulose and ash. However, the most significant changes in nutritional value take place as the plant grows. With growth the quantity of protein and ashes decreases, cellulose increases, and oil gets accumulates until the phase of bulb formation when its quantity decreases as well. By the end of blooming plants become harder and their forage value deteriorates noticeably. There is a decrease in digestible protein by 7–8.5 kg per 100 kg of dry forage, with a corresponding decline in fodder units. In the first stage of growth of *Artemisia terrae-albae,* the protein proportion is comparatively small 1:4.6–4.7. The nutritive value of 100 kg of *Artemisia* in winter pastures is 49.8–60.9 fodder units, or 3.5–6.4 kg of digestible protein.

Our research shows that certain components of desert grasslands are characterized by a high nutritive value, providing a good balance of digestible nutritious elements. All the ephemerons dry up in summer and the main ration of forage for animals is formed from *Artemisia*, *Ceratocarpus arenaria*,

TABLE 13. Chemical composition and nutritious value *Artemisia*-ephemeron grass stand of desert pastures in different years and seasons

Season	Samples	Year	Chemical composition				Protein content in 100 kg of fodder
			Water	Protein	Oil	Cellulose	
Spring	A[1]	1999	63.5	5.50	1.60	7.78	3.78
	A	2000	69.7	5.24	1.02	6.69	3.86
	B[2]	1999	64.0	6.02	2.02	8.05	3.98
	B	2000	68.2	5.58	1.50	6.31	4.13
Summer	A	1999	25.3	6.83	3.90	23.30	4.42
	A	2000	28.9	8.46	3.0	21.08	5.02
	B	1999	24.0	6.17	3.22	24.12	3.98
	B	2000	26.1	8.05	3.77	21.90	4.02
Autumn	A	1999	24.6	7.00	2.21	23.86	4.09
	A	2000	26.9	7.58	1.65	22.82	4.0
	B	1999	23.0	6.08	1.47	25.99	3.45
	B	2000	24.8	6.97	1.55	23.15	3.83

[1]Sample 'A' located at more than 5 km from settlements
[2]Sample 'B' located at various distances from settlements

dry ephemerons, while in autumn the principal components of the ration are *Artemisia, Ceratocarpus arenaria* and annual *Salsola*.

The method of pasture use directly affects the quality of forage. In spring time, when the vegetation is rich in feed value and fresh there is not much difference in the effects of grazing patterns and how they are used. In summer and autumn, however, it is possible to see quite a considerable difference on regenerated pastures. In summer in dry years, the total nutritive value of forage is increased by 5.5 fodder units and 0.44 kg of digestible protein per 100 kg of forage in comparison to pasture sites with unsystematic grazing. In wet years, the difference is more than 2.8 fodder units and 1.0 kg of digestible protein in 100 kg of forage, as is shown in Table 13.

6. Grazing impact on hydro-physical characteristics of soil

One of the main characteristics of soil is its weight. This indicator has a direct impact on the water, air and warmth regime of soils and consequently on its biological activity. Research shows that there were significant changes in the soil weights for all types of soils at our study sites (Table 14).

There is an especially obvious change in light sierozem type of soils. In 1999 in the immediate vicinity of Aydarly village, soil weight at 0–30 cm depth was 1.33 g/cm^3, while five years later it had increased to 1.35 g/cm^3. At increasing distances of 2.5 km and 5.0 km from livestock concentration points,

TABLE 14. Soil weight changes depending on intensity of grazing on pastures, g/cm^3

Samples, depth, cm	Desert zone Aydarly			Semi desert, Ulguly			Dry steppe zone Shien		
	Distance from villages, m								
	1000	2500	5000	1000	2500	5000	1000	2500	5000
Summer, 1999									
0–10	1.33	1.31	1.32	1.20	1.20	1.15	1.18	1.13	1.08
10–20	1.34	1.33	1.34	1.20	1.15	1.11	1.17	1.13	1.06
20–30	1.34	1.33	1.33	1.11	1.10	1.06	1.12	1.10	1.06
30–40	1.32	1.30	1.30	1.14	1.10	1.10	1.09	1.13	1.04
40–50	1.33	1.31	1.29	1.10	1.11	1.10	1.10	1.13	1.04
0–30	1.33	1.32	1.32	1.17	1.12	1.10	1.15	1.12	1.06
0–50	1.32	1.31	1.30	1.14	1.11	1.10	1.13	1.13	1.05
Summer, 2004									
0–10	1.35	1.33	1.34	1.24	1.21	1.17	1.20	1.15	1.08
10–20	1.34	1.35	1.35	1.20	1.15	1.11	1.18	1.13	1.06
20–30	1.35	1.34	1.32	1.14	1.12	1.08	1.12	1.10	1.06
30–40	1.33	1.30	1.30	1.16	1.15	1.14	1.10	1.04	1.04
40–50	1.33	1.30	1.29	1.10	1.10	1.11	1.11	1.04	1.04
0–30	1.35	1.34	1.34	1.19	1.16	1.12	1.17	1.13	1.07
0–50	1.34	1.32	1.32	1.17	1.15	1.12	1.14	1.13	1.06

the indicators of soil weight go down. Sierozem soils are also negatively affected by overgrazing, increasing indicators of soil weight by 0.01–0.02 g/ cm^3 in comparison to indicators that were measured in 1999. In dry steppe zones, chestnut type of soils also undergo compression. These changes could be easily observed in the upper layer of soil in 0–10 cm depth. Monitored sites that were lightly grazed did not have significant changes in soil weight. One conclusion we can draw is that the soil weight increase in many overgrazed places is caused by trampling since the most severely overgrazed pastures are along the livestock transit routes to pastures with forage. The problem here is livestock concentrations per unit of gazing territory which exceeds the permissible environmental load.

The maximum amount of dust in spring in 0–10 cm horizon (78.3%) occurred at a distance of one km from the village of Aydarly with light sierozem type of soils. Usually sierozem soils contain up to 65.3% of slit particles, while chestnut soils structurally contain up to 58.4%. As we go further from the village to a distance of five km, the soil samples contain larger particles, especially in dry steppe zones with chestnut type of soils. On the whole, large concentrations of livestock on a small territory lead to the formation of soil particles on the surface of which 58.5% are from 0.25 to 1.0 mm in size, which creates conditions for active wind erosion.

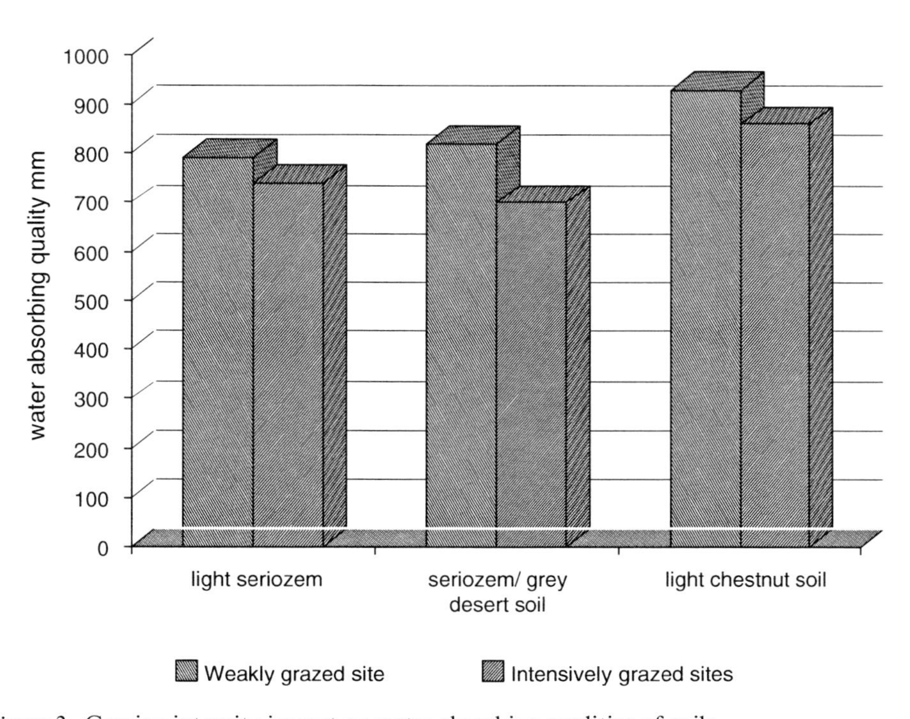

Figure 2. Grazing intensity impact on water absorbing qualities of soils

The water absorbing quality of soils also plays an important role in the water regime of soils and is the main factor to consider in terms of moisture accumulation and water availability for plants. The water absorbing qualities of soils are directly linked to soil structure and its compression (Figure 2). Figure 2 shows the significant influence of grazing on the water absorbing quality of soils at all sites.

Soil moisture indicators were measured over five years in all the monitored sites and revealed the following patterns: more moisture accumulation takes place in pastures coved by vegetation; places which are grazed less intensively preserve more soil moisture than pasture sites with intensive grazing. This trend is common across all the monitored sites (Table 15).

In sum, unsystematic grazing has a negative impact on the hydro-physical characteristics of different soil types causing deteriorating indicators of soil weight, structure, and moisture composition, all of which directly affects the formation and yield of forage on pastures.

TABLE 15. Soil humidity in the monitored sites along the transects, % of moisture in soil by weight (average for 2000–04)

Zone, village	seasons	Distance from villages, m	Sample depth, cm		
			0–30	0–50	0–100
Desert zone Aydarly	spring	1000	6.4	7.1	6.9
		2500	7.3	7.6	7.4
		5000	7.9	8.4	7.8
	summer	1000	2.9	3.6	3.9
		2500	3.4	3.7	3.9
		5000	3.7	3.8	3.9
	autumn	1000	3.4	4.2	5.0
		2500	4.1	5.0	5.5
		5000	4.8	5.3	5.1
Semi desert, Ulguly	spring	1000	17.1	17.2	13.8
		2500	17.4	17.1	14.2
		5000	18.6	17.8	14.8
	summer	1000	10.6	10.0	9.8
		2500	10.7	9.7	10.0
		5000	12.6	11.5	11.2
	autumn	1000	2.8	3.5	4.0
		2500	3.0	3.7	4.1
		5000	2.9	4.0	4.5
Dry steppe zone Shien	spring	1000	19.0	19.2	15.6
		2500	20.7	20.8	16.5
		5000	22.6	22.5	17.7
	summer	1000	18.1	16.8	15.0
		2500	20.1	18.5	15.8
		5000	21.2	20.0	17.2
	autumn	1000	1.3	1.5	1.7
		2500	1.1	1.3	1.6
		5000	1.5	1.4	1.7

7. Grazing impact on the chemical characteristics of soils

Soil samples were taken at nine study sties in order to analyze soil chemical composition with respect to grazing intensity. Most samples documented a decrease in soil productivity in 0–30 cm depth which mainly reflects humus levels. It is important to mention that up to 55% of the potential humus is lost in places with intensive grazing. This study also revealed that sierozem soils were very sensitive to overgrazing which caused a sharp decrease in humus composition in the upper layer of soil (Table 16).

Along with changes in the humus composition of the upper soil layers there are also changes in the amount of phosphorus with increasing grazing intensity causing a decline in the proportion of phosphorus in soils – up 2.61 milligram in chestnut type soils, 2.23 mg in sierozem type soils, and 2.14 in light sierozem type soils. A decrease in grazing intensity also contributes to phosphorus composition in the upper layers of the soil – up to 2.81 mg in chestnut soils, 2.0 mg in sierozem soils, and 2.74 in light sierozem soils. Some increase in calcium and magnesium composition could be observed in pasture sites with low grazing intensity.

8. Impact of pasture use methods upon sheep productivity

Between 1999–2004 we examined the impact of pasture use on the meat and wool production of sheep. Three groups of twenty-five animals each were grazed continuously around the village, and the other three groups each consisting of twenty-five animals were grazed on pastures used in rotation at a distance of five km from the village. Sheep were weighed in the morning before the animals were watered two times a year – once after lambing

TABLE 16. Changes in agro-chemical characteristics of soils in monitored sites along the transect (Aydarly village, desert zone, light sierozem type of soil)

Distance from villages, m	Depth cm	Humus, %	CO_2, %	Nutritious elements, mg /100 g of soil		Basic elements, mg /100 g of soil			Salt proportion, %
				P_2O_5	K_2O	Ca^{2+}	Mg^{2+}	Na^+	
	0–10	0.45	4.11	2.14	20.1	4.8	3.2	0.027	0.174
1000	10–20	0.54	5.04	2.83	21.4	5.6	4.8	0.032	0.200
	20–30	0.50	6.51	3.13	21.0	4.8	4.0	0.029	0.176
2500	0–10	0.66	3.19	2.40	22.0	6.0	2.8	0.034	0.212
	10–20	0.80	5.30	3.04	21.0	5.2	4.0	0.029	0.209
	20–30	0.63	5.96	3.23	22.0	6.4	4.0	0.089	0.172
5000	0–10	1.02	2.44	2.74	20.2	5.2	4.8	0.036	0.216
	10–20	0.60	3.59	3.14	20.3	5.2	3.2	0.029	0.192
	20–30	0.90	4.33	3.03	20.3	4.8	3.6	0.084	0.225

and again in autumn. Wool production was measured by weighing the fleece shorn from each animal. The monitored flocks were of the Kazakh fine-wool breed of sheep. Weighing results are shown in Table 18.

Table 17 shows that weight gains in sheep depend on pasture use and grazing methods regardless of the type of pastures that animals were grazed upon, although the pasture type did influence the rate of weight gain. Weather conditions (dry and hot versus wet years) have a direct influence on the body weights of animals. Weights were much higher in 2003 in all the monitored flocks when grazing sites were changed frequently. There was a positive change even in the weights of animals which were grazed in pastures around villages on account of very productive growth of ephemerons. Body weight dynamics is shown in Figure 3.

Figure 3 shows that maximum weight grain occurred in spring and to a lesser extent in autumn. Weights increased when livestock were grazed on regenerated pastures using a rotational system, giving an increase of 30% in comparison to sheep grazed continuously around villages.

TABLE 17. Sheep productivity under different methods of pasture use

			Pasture use method					
			Continuous grazing around the village			Grazing rotation at 5 km from village		
Type of pastures	years	Grazing period, days	Body weight of sheep, kg		Weight change, kg	Body weight of sheep, kg		Weight change, kg
Predominately grasses and *Artemisia* mixture	1999	220	30.2	38.4	8.2	30.7	45.4	14.7
	2000	230	32.0	40.5	8.5	31.8	44.1	12.3
	2001	220	30.0	37.9	7.9	31.4	43.6	12.2
	2002	217	31.6	38.1	6.5	31.9	45.3	13.4
	2003	226	31.2	37.2	6.0	31.4	48.4	17.0
	2004	230	32.2	40.6	8.4	31.7	47.8	16.1
Average					7.6			14.3
Artemisia-grass mixture	1999	226	29.9	38.6	8.3	30.2	46.0	15.8
	2000	234	31.3	40.1	8.8	30.6	45.4	14.8
	2001	220	28.6	37.8	8.2	29.3	44.0	14.7
	2002	222	30.8	39.6	8.8	30.2	45.1	14.9
	2003	231	31.4	40.4	9.0	31.1	46.8	15.5
	2004	241	31.6	40.2	8.6	32.0	44.9	12.9
Average					8.6			14.7
Artemisia-ephemeron	1999	238	30.8	40.0	9.2	31.6	47.3	15.7
	2000	240	31.0	38.9	1.9	31.2	47.5	16.3
	2001	247	31.2	39.4	8.2	30.7	46.0	15.3
	2002	234	30.0	37.7	7.7	30.3	45.1	14.8
	2003	230	31.8	38.9	7.1	31.1	47.6	16.5
	2004	241	29.3	40.9	10.6	30.1	48.8	18.7
Average					8.5			16.2

TABLE 18. Wool production of Kazakh fine wool sheep, females (average for 2004–06)

Livestock groups	Number of head	Management conditions	Wool shorn, kg (raw)	Scoured wool	
				%	Kg
1 group	25	Regenerated pastures with rotational system of grazing	4.4	55.71	2.45
2 group	25	Degraded pastures around the village	3.9	50.48	1.96

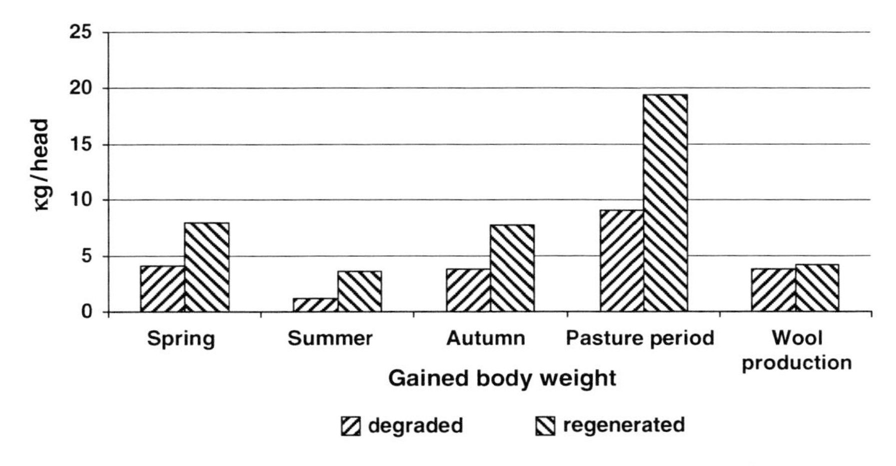

Figure 3. Body weight changes in degraded and regenerated *Artemisia*-ephemeron type of pastures

The system of pasture use also has an impact on wool production. Wool from female fine-wool sheep grazed rotationally on regenerated pastures had a fiber length of 4.3–4.6 cm, while that of sheep which had grazed around villages did not exceed 3.5–3.8 cm. Wool production data is summarized in Table 18.

Table 18 shows that wool production per sheep is increased by 0.5 kg and the proportion of scoured wool is improved by 5.23% if the animal is grazed rotationally on regenerated pastures, in comparison to sheep which are kept continuously on pastures around villages. Similar trends are evident from comparisons of Kazakh fine-wool sheep in *Artemisia*-grass mixture type of pastures in semi-desert and on pastures consisting predominately of grasses mixed with *Artemisia* in dry steppe zones. Raw wool production was 3.50 kg per sheep grazed on *Artemisia*-grass pastures while sheep produced 3.95 kg of raw wool when it was grazed on regenerated pastures of the same type. On regenerated pastures a female sheep produced on average 4.3 kg of raw wool, but only 3.9 kg from similar but degraded pastures.

In assessing the economic effectiveness of livestock management in pastures we have based our calculations on weight gains and wool production per head. This data was obtained from monitoring animals grazed on regenerated pastures using rotational grazing sites compared to animals kept on degraded pastures around Aydarly and Shien villages. Measurements were conducted on female Kazakh fine-wool sheep.

The results in this table demonstrate that a rotational systems of pasture use was the most profitable option on one private farm "Ospanov," which has 450 to 700 heads of animals. It is necessary to organize a pasture use system that takes into consideration the continuous seasonal rotation of pastures in order to avoid pasture degradation.

In sum it can be said that using regenerated pastures with rotational grazing is more economically effective than using degraded pastures around settlements caused by unsystematic use of pasture resources.

9. Conclusion

As Table 19 shows, the total value of sheep production on regenerated *Artemisia*-ephemeron pastures with seasonal rotation is 59.3% more than the value received from degraded pastures. For the time being, the present practices of pasture use to a certain extent prevent livestock raising on pastures from becoming a sustainable branch of livestock breeding. This is an important issue for the sheep and cattle breeding industries Kazakhstan and we should continue research to work out economically effective and environmentally safe strategies for using pasture resources.

One of the radical approaches to pasture use in south-east Kazakhstan is mobile livestock breeding taking into a consideration the present principles of land use and land ownership. Mobile livestock breeding based on the regular rotation of seasonal grazing sites provides for the normal growth and evolution of pasture vegetation and the productive longevity of pastures even in year with minimum rainfall (such as 2001–2002). Experiments conducted between 1999–2004 on six different types of pastures in this region demonstrated the negative impact of unsystematic grazing with livestock concentrated in one place, which resulted in low animal performance. Output can be improved when livestock are grazed seasonally on different types of pastures taking into account the carrying capacity of each pasture type.

As an overall result of our research and based on theoretical considerations we have developed a scheme of pasture use for different types of pastures located in the south-east regions of Kazakhstan. The main principles of this scheme require adherence to the following recommendations:

TABLE 19. Economic effectiveness of sheep production depending on *Artemisia*-ephemeron pastures under alternative management systems

Indicators	Measurement units	Sheep grazing sites: Degraded pastures around villages	Sheep grazing sites: Regenerated pastures with rotational system of grazing	Effectiveness indicators (±)
Female sheep, number	Head	25	25	-
Wool shorn (raw)	Kg	3.9	4.4	+0.5
Sale price per kg	tenge[1]	250	250	-
Total value of wool production	tenge	24375.0	27500.0	+3125
Body weight gained per head	kg	8.3	15.5	+7.2
Total body weight gained by flock	kg	207.5	387.5	+180
Total cost of body weight gained	tenge	41500.0	77500.0	+36000
Total value of meat production	tenge	65875.0	105000.0	+ 39125
Expenditures for winter fodder, per head of animal per a day:	tenge			
2 kg Lucerne hay		16.0	16.0	-
0,2 kg wheat-forage		5.0	5.0	-
Costs per head for 100 days	tenge	2100	2100	-
Winter fodder cost, total	tenge	52500	52500	-
Income	tenge	13375.0	52500.0	+39125

[1]122 tenge = $1 USD

- Regenerated pastures should be used seasonally by rotating grazing sites. Using regenerated mixed *Artemisia*-grass pastures without rotation will lead to pasture degradation.
- Since at the present time it is often impractical to move livestock from one pasture to another which is located 100 km or more away, rotational grazing sites within the one seasonal pasture need to be developed. A biological knowledge of these vegetation communities should be a basis for this work. The general lesson repeatedly demonstrated by research work recommends that ephemeron-ephemerid pastures be used in spring

time, that those with grasses be used in summer, that *Salsola*-dominated pastures be used in autumn, and that pastures consisting primarily of sub-shrubs are good for winter forage.

- As regards the appropriate number of times a pastures should be used in a year, we have come to the conclusion that all the pasture sites should be used once only in the season. Only in favorable years, such as 2006, is there a possibility of using the same grazing sites two times in a single year, if there are good ephemerons and strong growth of summer vegetation.

Acknowledgements

Support for this research was provided by the European Commission Inco-Copernicus RTD Project ICA2-CT-2000-10015 'Desertification and Regeneration: Modeling the Impact of Market Reform on Central Asian Rangeland' (DARCA), and under a United States National Science Foundation Grant No. DEB-0119618 'Biocomplexity, Spatial Scale and Fragmentation.'

References

Laryn, I.V., 1956, Meadow Cultivation and Pasture Farming, Moscow and Leningrad. [in Russian].

Zhambakin, Z. A., 1995, Pastbisha Kazakhstana [Pastures of Kazakhstan], Almaty: Kainar [in Russian].

CHAPTER 6

LIVESTOCK MOBILITY AND DEGRADATION IN KAZAKHSTAN'S SEMI-ARID RANGELANDS

SCALE OF LIVESTOCK MOBILITY IN KAZAKHSTAN

CAROL KERVEN*[1], KANAT SHANBAEV[2], ILYA ALIMAEV[2], AIDOS SMAILOV[3] AND KANAT SMAILOV[2]

[1] *Macaulay Institute, Craigiebuckler, Aberdeen AB15 8QH, UK*
[2] *Kazakh Scientific Centre for Livestock and Veterinary Research, Dzandosov Str. 31, 480035 Almaty, Kazakhstan*
[3] *Association of Oil and Gas Energy Sector in Kazakhstan, Almaty, Kazakhstan*

Abstract: Kazakh pastoralists formerly followed long-distance migratory routes each season. This was continued with state farm support during the Soviet period. After the collapse of state farms in the mid 1990s, most pastoralists were constrained to graze their animals in circuits around villages, as they could not afford to undertake seasonal migrations. Pasture degradation has resulted. Small-scale village-based livestock owners rely on their animals mainly for subsistence. Compared to large-scale owners, they gain higher rates of economic returns per head of animal owned. Large-scale owners have returned to moving their animals to distant pastures, and their animals are heavier as a consequence. These types of owners can achieve economies of scale, but they have high actual costs of moving animals.

Keywords: livestock mobility, pastoralists, household economy, Kazakhstan

1. Introduction

Kazakh nomads practiced seasonal mobility on a large geographical scale as the traditional form of animal husbandry in the pre-Soviet period (Zhambakin, 1995; Olcott, 1995). Livestock mobility was interrupted or

* To whom correspondence should be addressed. Carol Kerven, Macaulay Institute, Craigiebuckler, Aberdeen AB15 8QH, UK; e-mail: carol_kerven@msn.com

R. Behnke (ed.), The Socio-Economic Causes and Consequences of Desertification in Central Asia, 113–140.

compressed during the Soviet period, virtually ceased in the immediate post-Soviet 1990s, but is currently re-emerging as flock numbers rebound from the mid 1990s livestock population crash (Behnke, 2003; Kerven et al., 2006). In other semi-arid regions of the world, reduction in the extent of livestock mobility has been associated with alterations in rangeland vegetation and soils (Behnke and Scoones, 1993; Galvin et al., 2007; Humphrey and Sneath, 1999). Reduction of mobility, as livestock are concentrated around settlements, has previously been found to lead to rangeland degradation in south eastern Kazakhstan (Alimaev, 2003; Alimaev et al. this volume; Alimaev and Behnke, 2007; Ellis and Lee, 2003). This paper presents data on some of the socio-economic causes and consequences of contemporary rangeland mobility or immobility by pastoralists in the same southeast regions of Kazakhstan.

Over the past century, Kazakh livestock have been moved at scales which have expanded and contracted according to political and economic upheavals. These external forces on the livestock system resulted in a loss of access to forage and water resources that were previously contained within unfragmented migratory cycles. The initial fragmentation of Kazakh grazing grounds came about in the 19th century from an intrusion of Russian cultivators, backed up with military power (Martin, 2001). Later there was an enforced sedenterisation of nomads (Olcott, 1981), and lastly the dissolution of state-sponsored farms (Behnke, 2003; Kerven, 2003). But only in the first instance of settler agriculture encroaching onto the steppes did the ranges become physically separated from each other or excised from grazing. In the latter two time periods, livestock were drawn back from their extensive pastures, which became unused but remained contiguous, unfenced and communal - in other words, unfragmented. It was the migratory cycles which were broken apart as new production systems were formulated. The fragmentation of livestock grazing areas is not necessarily caused by physical barriers, or the physical change of land tenure systems, but can result from socio-economic and political changes.

Rangeland degradation in Kazakhstan is directly affected by whether animals are moved to seasonal pastures or are grazed all year around settlements and permanent water points. This was shown by previous research from 1998–2002 (Kerven et al., 2003) and confirmed by recent research conducted for the SCALE[1] project from 2004 to 2006 (Alimaev

[1] National Science Foundation (USA) BE/CNH: "Biocomplexity, Spatial Scale and Fragmentation: Implications for Arid and Semi-Arid Ecosystems" Award Number: 0119618, managed by Natural Resources Ecology Lab, Colorado State University, Fort Collins, Colorado, USA.

et al. this volume). Other factors being equal, the scale of livestock movement was founded on the temporal and spatial variability of Kazakhstan's rangeland ecology (Asanov et al., 1992; Zhambakin, 1995). This often mirrored that of the wild saiga antelope which migrated across the same ranges (Bekenov et al., 1998).

"Underlying this system of seasonal pasturage was the dictate that grasses remain untrammelled and uneaten during the seasons not designated by the nomadic group for its use. Thus in the words of Chokan Valikhanov ... (1835–65), the Kazakh ethnographer and imperial official 'while it appears that the Kyrgyz [former term for Kazakhs] use enormous expanses of land, in fact they only use a little at a time'" (Martin, 2001: 20).

Historically however, the shifts in political and socio-economic conditions for Kazakh pastoralists have meant that other factors have rarely been equal, and nomads have not always been able to pursue their seasonally mobile grazing strategies (Alimaev and Behnke, 2007).

The scale of movement is currently dependent on the social and economic costs of movement which must be bourn by the individual pastoralist family. In turn, these costs are closely associated with the scale of individual livestock ownership; owners of bigger flocks move their animals across much larger areas of land which is more ecologically heterogeneous, because the owners possess the social and financial capital required to move their animals seasonally.

These 21st century bigger-scale owners have partially and consciously reconstituted the old migratory cycles of their nomad forbearers in the 19th and early 20th century. Soon after the collapse of the state livestock farms in the mid 1990s, very few pastoralists had many private animals and almost all had lost their jobs on the state farms (Behnke, 2003; Kerven, 2003). In 1998 an emerging large-scale pastoralist was interviewed in one of the project research sites. Through having privatised a local garage formerly owned by the state, and illegally selling firewood from the desert, he had quickly accumulated a considerable number of private livestock for that time: 200 sheep, and 140 cattle and horses. He asserted:

"My grandfather was a *biy*[2], a very wealthy man, but in 1929 their livestock were taken by the government. Before that, Kazakhs could keep their animals by moving to distant pastures; my grandfather had vast areas of pasture under his control and he knew what to do with this land. I would like to get 100,000 ha of pasture and bring workers there to herd my animals, and the herders

[2] In Kazakh nomadic institutions, a *biy* primarily functioned as a judge but some acceded to this position through the livestock wealth of their kinship group (Martin 2001). After the imposition of Bolshevik rule, the rich *biy* (lords) who owned many thousand head of livestock, controlled grazing land, water points and relied upon clan-based herding labour, came under attack and were either killed or fled to other countries in the Stalinist period (Olcott, 1995).

would get income. People need jobs. I would not need to fence the pastures because in the past people divided up their pastures without fencing. Not all people are equal, as five fingers on a hand are not the same. I should be able to inherit the position of *biy*".

The difference between this past and the present is that the extensive scales of seasonal movements in the nomadic past were enabled through kinship membership in clans governed and protected by tribal and military leaders (Martin, 2001). Long distance livestock movement across unfragmented rangelands is still entirely possible and is being practiced, but only by a minority of livestock owners who must rely on their own much more limited resources. The former nomadic social scale that allowed vast distances to be travelled to gain access to seasonal pasture and water, was replaced in the 1940s by the highly mechanised and subsidised state farm (*sovkhoz*) system which operated at a huge scale, owning 60,000 sheep and thousands of cattle, horses and camels. The *sovkhoz* provided the institutional and material support that long distance migration required. At the present time, the social scale of the mobile unit has now shrunk to small patrilineal groups of fathers, brothers, nephews and sons managing large flocks with non-related hired shepherds. These historical patterns imply a symmetry between institutional and spatial scale in managing rangelands, a point acutely observed by Humphrey and Sneath (1999) on the scale and fragmentation of Mongolian pastoral migrations.

2. Rationale for studying pasture use and livestock management

The SCALE study in Kazakhstan sought to measure the biological and economic returns to seasonal livestock movement compared to circum-village and sedentary grazing management. The aim was to determine the costs and benefits to livestock owners of moving their livestock to seasonal pastures. This strategy is compared to having animals forage in one location throughout the year, with supplementary feeding in winter as necessary. Another component of the SCALE study took vegetation samples at the seasonal grazing sites used by the livestock belonging to the sample of households. The results are reported in Alimaev et al. this volume.

The Kazakh government is defining new policies towards livestock and rangeland management, as revenues from the booming mineral economy are released for rural reconstruction and agricultural research. Major international donors such as the World Bank (2003; 2004) and UNDP (2007) have designed rangeland improvement programmes to rehabilitate land that was previously overgrazed by large numbers of state-owned livestock, or poorly irrigated, or mechanically cultivated. Unfortunately, these projects may be designed without objective or current information from field research on how Kazakh pastoralists are managing the rangelands in the post-Soviet period. Nor are the underlying social, institutional and economic causes of new patterns of rangeland degradation necessarily considered. Instead, these projects assume that rangeland users – the pastoralists - have "insufficient knowledge" of pasture degradation, range ecology and optimal land management (UNDP, 2007).

Meanwhile the Kazakh government focus of livestock development is mainly on specialised breed improvements for the benefit of larger scale herd and flock owners. Policies are yet to be defined on the optimal use of the nation's extensive rangelands, comprising 60% of the land area and amounting to 189 million hectares. Though most of this land receives less than 300 mm of precipitation a year and is unsuited for cultivation, these dry and remote rangelands are yielding Kazakhstan's great wealth of oil and gas. There is less recognition of their ecological value in supporting grazing livestock, biodiversity, carbon sequestration and other benefits including cultural values.

Since the mid 1990s, entirely new systems of private livestock and rangeland management have spontaneously developed, about which there is little information. There is a need for contemporary and relevant data to guide policies on extensive livestock management. One of the policy issues is whether sedentary smaller flocks that are in the majority can be as productive as the large, mobile flocks, whose owners are often well-connected politically and receive most attention from policy-makers and state researchers who must rely on government funding. For example, new government subsidies are aimed at large-scale livestock owners with more than 1,000 head of sheep, as the government perceives this group to be more productive and progressive.

The question of relative productivity in large mobile flocks versus small sedentary flocks arises from previous research (DARCA[3]) conducted from 2001 to 2004 in two different ecological regions of southeast Kazakstan, on the pattern and scale of livestock movements in relation to forage resources, livestock productivity, and economic returns to the livestock-owners. The two regions were formerly intact - or unfragmented - livestock migratory cycles at the end of the Soviet period in 1990, as described in Alimaev and Behnke (2007).

In the previous DARCA research project, in a stratified sample of 46 livestock owners in six villages across two regions, the majority of livestock owners had less than 100 head, mainly sheep and goats. Most of these small flock owners grazed their animals around the villages all year, or in cold winter periods, stall fed in barns. A minority of owners either moved their animals in each of the four seasons, or else kept their animals at outlying barns up

[3] DARCA was the "Desertification and Regeneration in Central Asia: Modelling the Impacts of market reforms", funded by the European Commission INCO DEV, and managed by The Macaulay Land Use Research Institute, Aberdeen, UK.

to 40 km from the villages. Several thousand sheep and goats belonging to the sample households were weighed every three months for two and a half years. The results showed the effect of seasonal movement on sheep weights, particularly from winter to spring. Sheep which were moved in each of the four seasons gained on average 5 kg weight over winter, compared to village-based sheep which lost on average 8 kg, being stall fed or foraging on over-grazed ranges within 5 km of villages (Kerven et al., 2003; 2006).

There may be contradictory biological and economic indices of the relative success or attractiveness of alternative production systems. Seasonally moving animals to temporary pastures may produce heavier animals, but at what cost to their owners? The previous DARCA research found that mobile flocks had higher costs per head of animal, due to the need for mechanised transport and hired shepherding labour. Are heavier animals more financially rewarding to their owners over the longer term? What scale of land use is most profitable to livestock owners who are themselves at different economic scales? If loss of mobility results in increased land degradation around settlements and water points, what is the economic benefit? Providing some answers to these questions was the impetus for the small follow-up SCALE field study in two of the six former DARCA study areas in Kazakhistan.

3. Study areas

For this SCALE study, two villages that had been included in the previous DARCA project were selected along with their grazing orbits and long distance migratory circuits. The rangeland vegetation and effects of grazing at different scales is described in more detail in Alimaev et al. (this volume). The first study village is Ulan Bel, in the desert zone of Moinkum district in Jambul *Oblast* (province) in the south central part of the country. Average annual precipitation is 142 mm (ranging from 124–167 mm between 2003 and 2006) and temperatures are extreme, falling to −42°C in winter with a maximum of 47°C in summer. The village of Ulan Bel is located along the Chu river which runs through the desert, and the reeds around the Chu flood plain are a crucial livestock winter feed resource in an otherwise arid climate. The vegetation of the migratory grazing circuit varies. To the south of the Chu river lies the Moinkum desert (the name in Kazakh means "sand up to the neck"). This area of sand dunes is dominated by shrubs (*Haloxylon persicum*, *Kochia prostrata*, *Calligonum* and other species). The riverine area contains reeds (*Phragmites* and *Tamarix*), and a large spring flood plain with *Agropyron, Festucca* and *Artemisia* plant species. North of the village lies the plain of Betpak Dalla (in Kazakh, "the ill-fated steppe, also known in Russian as "the hungry steppe") poorly vegetated with saltworts, *Artemisia* and *Salsola* species. Beyond Betpak Dalla plains, more than 200 km to the north, lies the Sary Arka ("yellow band") of low mountains and steppe grasses. This was formerly used by nomads and later by collective farm shepherds for grazing animals over summer. Ulan Bel village had been the centre of an important state livestock breeding and production farm supplying pedigree karakul sheep to other desert farms in the region from 1948 until 1998, when it became bankrupt. As most people in and around Ulan Bel now have no formal employment, hunting or poaching for wolves and saiga antelope in the Betpak Dalla have become important sources of income, in addition to collecting and selling fallen rocket parts from the Baikhanour rocket launch pad. The population had halved to around 1,400 and many houses as well as public buildings were in ruin. A defining feature of the desert villages is their remoteness; Ulan Bel is 220 km from the nearest city of Taraz, a 10 hour roundtrip drive on very poor roads.

The second study village is Aidarly, lying on the edge (Kazakh: *jeek*) between the desert-steppe to the south, and the Sarytaukum ("yellow mountain of sand") desert to the north. Aidarly lies in the northern part of Jambul district of Almaty province, in the south east of the country. The average annual precipitation is 230 mm, while temperatures range from a minimum −35°C in January to a maximum of 45°C in July. Vegetation in the northern sand dune area of the Sarytaukum is mainly composed of shrubs, *Haloxylon persicum*, *Artemisia* species, small shrubs of *Salsola* and *Kochia*, and in spring, ephemeral grasses of which *Carex physodes* is the most important. Some 20 km south of the village lie some low hills with springs, dominated by *Artemisia* species and ephemeral spring grasses.

Aidarly village is 200 km from the nearest urban centre of Almaty, 6 hours roundtrip driving on good roads. The village was formerly the administrative

Plate 1: Mountain summer pastures of Ala Tau, formerly used by Aidarly pastoralists. All photographs in this chapter by Carol Kerven

and residential centre of a state farm (*sovkhoz*) keeping sheep from 1960 to its dissolution between 1993–95. As jobs disappeared, many families left the village, but a number of ethnic Kazakh families repatriated from Mongolia and China have settled in the village since then, and two were included in our sample of families. The population is about 1,000, but many houses and public buildings had collapsed. During the Soviet period, livestock belonging to the state farm were taken in summer to graze for three months in the Ala Tau sub-alpine mountain meadows (Plate 1). Since the end of the state farm, no private livestock owners from Aidarly have undertaken this long distance summer migration of several hundred km.

4. Methods

The SCALE study covered two years, starting in autumn 2004 and ending in winter 2006. Twelve flocks and households were monitored over the two years, in the two study areas of Ulan Bel and Aidarly, described above. These represent different rangeland ecologies in south eastern and central Kazakhstan. In each region, one village and its total grazing environs was selected, which had previously been in the larger DARCA study sample. Within each selected village, six cooperating livestock owners and a sample of their sheep and goats were monitored. In each village the selected owners comprised two who grazed their animals all year around the village, two keeping livestock year-round at outlying barns more than 20 km from the village, and two owners who moved their livestock each season to different pasture zones up to 200 km distant from villages. In sum, twelve flocks with a total sample of approximately 400 sheep and goats were monitored over two years, giving 24 flock and household data sets. Flocks were selected from the major breed kept in each ecological zone. Owners keep several breeds of sheep and goats, including crossbred animals.

Live weight changes in sheep and goats were measured twice a year, once in spring (March and May) and once in autumn (October and November). From the previous research in the study area, it was known that the main weight changes occur over winter when grazing is restricted and cultivated or cut fodder may have to be provided, and between spring and autumn when animals gain weight on green pastures. For each sampled flock-owner, up to 40 ear-tagged sample adult female animals were weighed, representative of age classes within each flock. Weighing was carried out in the animal pens, either early in the morning before the animals were let out, or in the evening after returning. Progeny histories were planned to be recorded on the ear-tagged adult females, to assess lambing/kidding rates per female and lamb/kid survival to weaning, but this data could not be analysed as flock owners were not able or willing to keep complete records over the two year period.

An economic questionnaire was applied twice to each sample household, once in November 2004 and a second time in March 2006. The questionnaire obtained data for the previous 12 month period on:

- Number of livestock owned and managed, by breed
- Capital equipment owned and used for livestock, e.g. trucks, barns, well pumps
- Live animal and livestock product sales and prices obtained
- Livestock purchases
- Home slaughter rate
- Transport and labour costs incurred in managing livestock
- Costs for obtaining winter supplementary feed, and veterinary costs
- Amount and type of fodder given in winter, by animal category
- Other income from crops, employment and government benefits

In addition, each sampled household was informally interviewed in-depth once and sometimes twice a year, to record their grazing patterns in each of the four seasons, including water points and pastures used, as well as length of time animals spent in each location and winter fodder regimes. These interviews were semi-structured, and revealed the opinions of the flock-owners on the issues of livestock productivity, winter feeding, pasture use and seasonal movement.

There are methodological advantages and disadvantages to a very small sample of households. The advantages are that in-depth interviews could be conducted several times with each household over the two year period, on household economy, livestock production and movement patterns. Each household was known to the researchers, who stayed with some of the families or with the village veterinarian, a very useful source of information about the circumstances of each family. In analysing the questionnaire survey results, the researchers could interpret the data according to their prior knowledge of each family. For example, when a poorer household had sold a lot of animals in one year, it was known that a son of this household got married in that year, and had to provide the customary brideprice and feasts. A further advantage is that only one researcher asked the survey questions each time, and he became trusted by the pastoral households in the survey.

The disadvantage of the small sample size is the difficulty of drawing statistically valid conclusions that could be confidently applied to the pastoralist population at large in the study areas. With only twelve families in the sample, the survey results will considered in conjunction with illustrations gained through the in-depth interviews, on a case study basis.

5. Survey findings

The sampled families were all mainly dependent on livestock for their income. Numerically, sheep are the main species kept, followed by goats, cattle, then horses and finally a few camels. Among the twelve families, only two grew some crops and only in one out of the two study years. Only two families had an employed member in one year, and only one family in the other year. Four families received some social security payments (pensions) in one year, and six (half) the families in the next year (2006). Only one family harvested their own hay as winter feed for their animals; all the rest purchased fodder for their livestock in both years. Co-residential family sizes are small, typically a husband, wife and several children, often a married son and daughter-in-law. The largest flock owners had from one to four hired herders who camped with the owners' family on migrations. All but the poorest families kept at least one milk cow and one riding horse used for shepherding (Plate 2).

Plate 2: Shepherds on horse back

The costs and returns to different herding strategies over the two year period are shown below in Figures 1, 2 and 3. These calculations do not include the opportunity cost of family labour, which is low for the less educated pastoralists in the desert regions of this sample. The net income also does not factor in the reproductive increase of the flocks, in terms of animals surviving to weaned age (which could not be recorded), or the value of cows' milk consumed by the family, also not recorded.

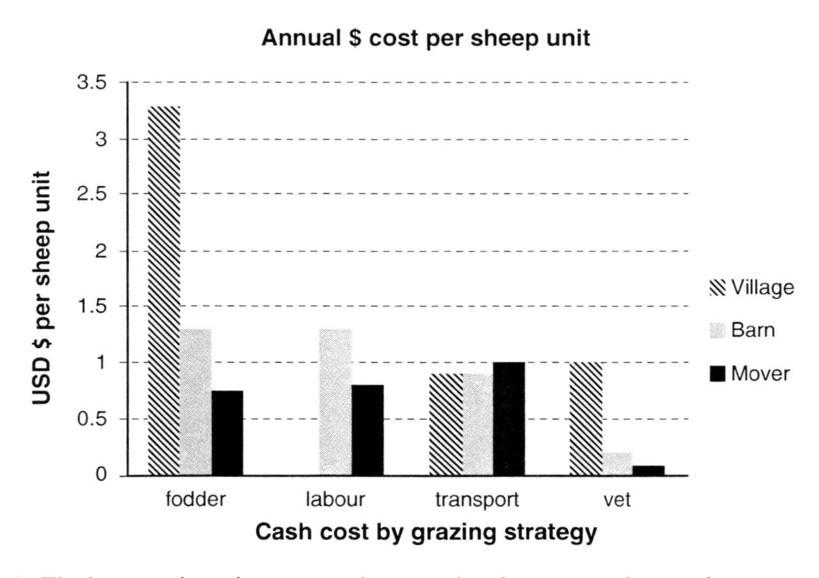

Figure 1. Flock owners' grazing strategy by annual cash costs per sheep unit

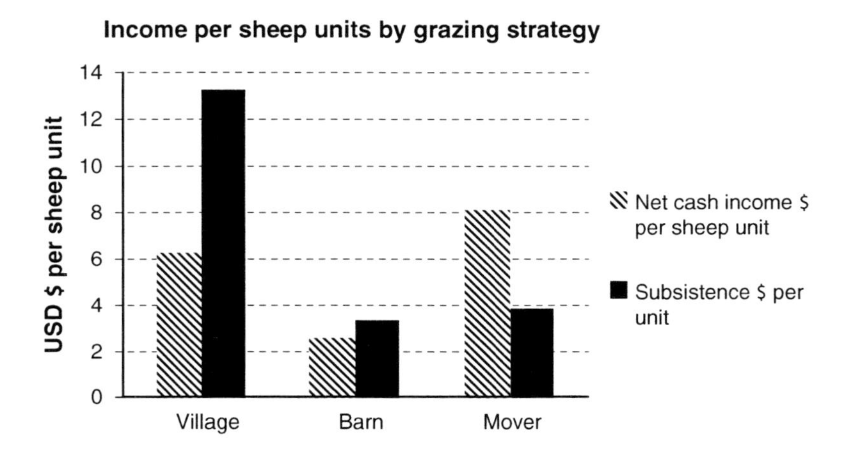

Figure 2. Flock owners' grazing strategy by annual net returns per sheep

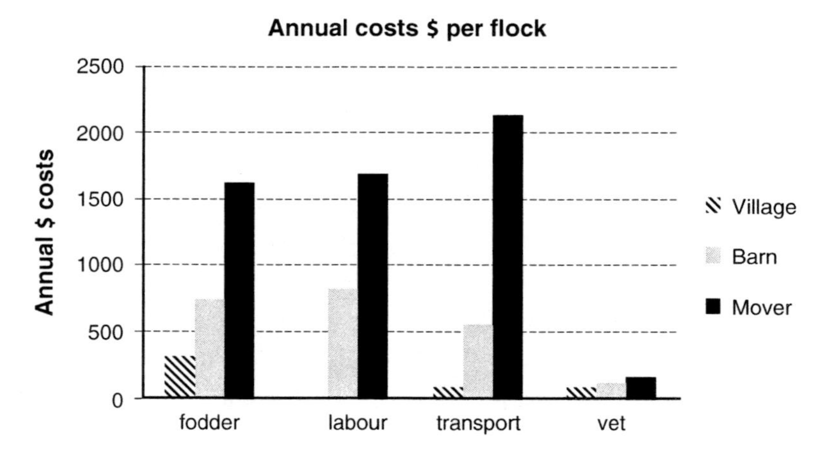

Figure 3. Flock owners' grazing strategy by annual cash costs per flock

The results were calibrated according to sheep units, since all families owned several species of livestock. Sheep units were calculated based on a metabolic live weight of a 45 kg sheep. Based on the DARCA data for weighed sheep and goats in both villages, the Ulan Bel sheep unit was 1.21 and goat 0.98, while Aidarly sheep unit was 1.19 and a goat 0.95. The ratio for a cow was 3.61, horse 4.66 and camel 5.62 based on estimated live weights from the literature.

The results show that village-based sedentary families with small flocks of less than 100 sheep units (range 50–200) had the highest cash costs of $5.9 per sheep unit per year, mostly on winter fodder (Figure 1). However, village-based families also had the highest overall net returns per sheep unit $19, when the equivalent market value of the animals consumed by the family was included (Figure 2). Home consumption of animals by the village-based families ranged from 3 to 18 sheep units per year.

The barn-based families had annual cash costs of $3.3 and net returns of $6 per sheep unit (Figure 2). They managed on average 660 sheep units (range 440–830), mostly sheep with some goats, cows and horses. These were mainly family operations, not heavily reliant on hired herders. Compared to village-based families, their consumption of sheep units was slightly higher.

The migrating families had annual cash costs of $3.7 per sheep unit and gained an average of $12 in annual net returns per unit (Figure 2). They owned on average 1,300 sheep units, ranging from 800 to 1,880, which included some camels as well as goats, cows and horses. However, the net returns include the value of up to 80 sheep units consumed per year, due to feeding hired herders who eat with the owner's family, and this should be considered as an operational cost rather than a return.

What these results show is that small sedentary flock owners have to buy relatively more fodder to keep each of their animals alive over winter (Figure 1), but their flocks are kept mainly as a source of food for the family and cash when sold at the markets (Figure 2). For individual village-based families, the annual value of animals consumed is always at least equal and up to double the cash value from selling their animals, wool, cashmere and skins, which averages USD $560 per year. The village sample did not own items of capital equipment for livestock management, except horse carts or small saloon cars. Figure 1 also shows that villagers spend proportionately much more on veterinary costs per head of livestock, but spend nothing on hired labour, in contrast to the mobile flock owners.

The barn-based flock owners lack the large numbers of animals to justify seasonal movement, with half the average flock size of the large-scale owners; they also lack the financial and capital resources which are necessary to undertake seasonal movements. Their costs and returns lie in between those of the village-based and migratory flock owners. Their annual cash income from livestock averaged USD $ 1,700, ranging from a loss of USD $1,100 to a maximum of $4,400.

The largest flock owners are all migratory, and though they have high flock costs for labour and transport (Figure 3), they do not have to buy nearly as much fodder per head of livestock over winter, compared to village-based animals, as their animals are taken to winter grazing grounds. Their net cash returns are accordingly higher per animal managed (Figure 2). This group all own many items of heavy mechanical equipment – large Russian trucks, tractors, hay balers, water pumps, water tanks, wagons used by hired shepherds, traditional Kazakh felt tents (yurts), and motor cars (Plate 3). These large flock owners are able to take advantage of economies of scale. Their unit costs per animal remain fairly flat for capital equipment (trucks, tractors, water pumps, wagons etc.) and labour, while they can greatly reduce the variable costs of purchased fodder per head, by taking their livestock to the remote grazing areas for winter.

Although transport is the single biggest cost for mobile flocks (Figure 3), when the transport costs are pro-rated per sheep unit, (Figure 1) the economy of scale in a large flock becomes apparent – each head of livestock accrues a not dissimilar transport cost to a village-based livestock head. Villagers have to spend cash on transporting their fodder and water for their livestock in the barns over winter. For mobile flock owners, annual net cash income from livestock averages $ 8,240, (ranging from a loss of $1,000 to a maximum of $16,000) almost entirely from selling prime aged male animals to the large city markets.

Results from the autumn 2004 and spring 2005 animal weighing surveys indicated that the sheep moved seasonally to graze on the most distant and

Plate 3: Camels and tractors owned by large-scale pastoralists

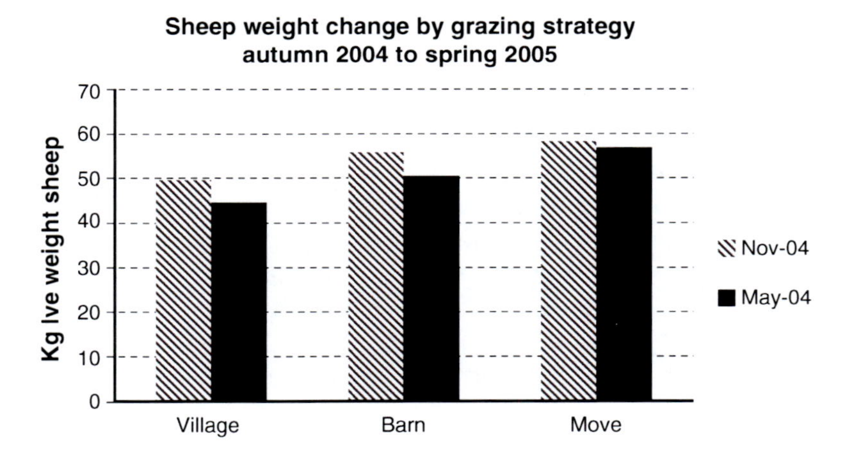

Figure 4. Flock owners' grazing strategy by sheep weight changes 2004–2005

most productive pastures did not lose weight over winter (Figure 4 and Plate 4). Pastures in the winter grazing grounds yielded two to three times the dry matter of peri-village pastures in the Moinkum study area (Table 1). The village-based sheep grazed around the villages in three seasons and partly housed over winter with supplementary feed lost between 4 to 6.5 kg live

TABLE 1. Yield of dry fodder mass at sampled grazing sites, kg/ha mean for 2004–2006 Ulan Bel sites in Moinkum

	Kg/ha by season			
Pasture sites of grazing flocks	Spring (April)	Summer (July)	Autumn (Oct)	Winter (Jan)
Village, up to 5 km radius	140	210	160	110
Barn, up to 5 km radius	210	260	230	180
Mobile	310	430	370	320

Plate 4: Winter grazing in the sand dunes

weight between November 2004 and May 2005. Moreover, village-based sheep weighed considerably less, by 6 to 14 kg, going into winter, compared to sheep of the same breeds that were moved each season in 2004 to distant pastures.

Weight results from the following winter and spring season (2005/2006) were less clear (Figure 5), mainly because the 2006 spring weighing took place in very early spring, when new vegetation was not available for grazing, compared to May 2005 in the previous year when migratory livestock kept away from the villages on the distant pastures could access new growth.

There were no significant differences in weights of local goats between winter and spring, in either year. Goats, being less selective foragers, seem to be able to maintain their weight over winter even when kept around villages. Poorer villagers also concentrate more on raising goats rather than sheep, as local goats can kid twice in a year, and often produce twins. Since goats are also less likely than sheep to lose weight over winter if not given more expensive fodder, this is another reason why they are preferred by poorer families (Plate 5).

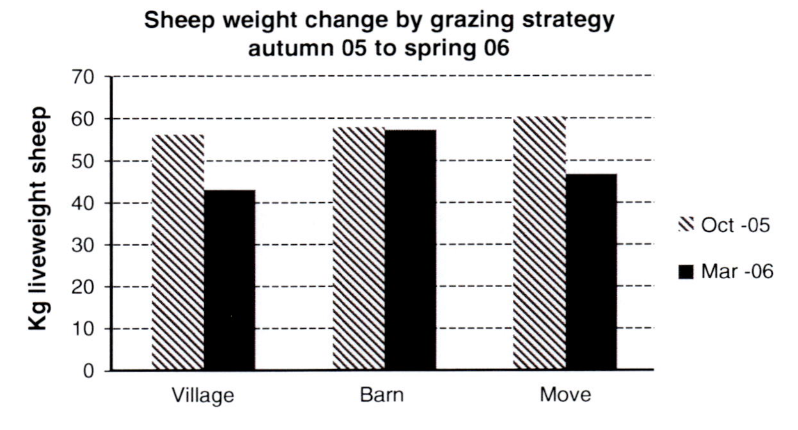

Figure 5. Flock owner's grazing strategy by sheep weight changes 2005–2006

Plate 5: Goats at a barn 40 km from Aidarly village

6. Owners' flock management strategies

These quantitative data results support the qualitative findings from in-depth interviews. Small flock owners have to adopt different flock management strategies than large flock owners, and to that extent, their economic returns are not strictly comparable. Poorer families have to graze their animals around the villages as they lack social capital in the form of resident adult male kin: older sons, nephews and brothers; the physical capital in the form of trucks, barns, pumps for wells etc., and financial capital to pay wages to hired shepherds and purchase fuel for vehicles and pumps. Following are some examples of small flock owners' strategies for their flock management. Real names are not used (Boxes 1 and 2).

BOX 1. Small flock owners: two brothers

Daulet and his older married brother Jerken lived in the ruined and nearly deserted village of Mali Kamkale, in the administration of Ulan Bel. In 2001 they had 15 sheep, 10 goats, a cow and a calf. Daulet had formerly been employed as a tractor driver in the sovkhoz (state livestock farm), which became bankrupt in 1998 and he lost his job. Then his brother Jerken moved to Kazakhstan's main city of Almaty, and worked in a factory but he returned in 1999 to help his brother Daulet manage the animals after their parents died. Both brothers planned to move to Almaty, but said "If conditions improve here we will stay as this is our native land". They had sold 10 cows, 100 sheep and a couple of horses to implement their plan for moving to Almaty. In autumn 2000 they bought locally 1.5 tonnes of hay made from reeds around the Chu river, and 1 tonne of mixed barley and wheat bran bought from the city of Taraz which is 10 hours driving each way. They started feeding their animals at the end of December and all the feed was finished by the end of February, though winter continued and they remarked that there was no pasture around the village in winter. Twice a day they fed a total of 8 kg of barley and wheat, plus hay, to the sheep and goats, and the cow received 2 kg per day, in addition to reed hay. Daulet said "It's important to give extra feed to animals in winter, as they are pregnant".

BOX 2. Small flock owner: woman and her goats

Sveta lived in the village of Ulan Bel, with her husband and young daughter. Both she and her husband were unemployed, and he got money by accompanying hunters for wolves and saiga antelope, in the Betpak Dalla plains. In May 2004 Sveta had 30 goats and 6 sheep, with no cows or horses. They grew no crops, received no government support and owned a motor bike but no vehicle. Her animals were managed in a neighbourhood herding group (Kazakh; *kyzk kyzk*), composed of three families who took turns providing a family member to take their combined animals out to graze in a 5 km radius around the village every day. In the autumn of 2003, her family harvested reeds near the Chu river, collecting 1.5 tonnes to feed her animals over the winter, and transported the reeds by motor bike to her village home. Sveta also went to the city of Taraz in the autumn, a 10 hour round trip, to buy barley as a concentrated feed for her animals over winter. She hired a taxi to bring the barley back to Ulan Bel, where she said no barley was available to buy. Over the winter of 2003/2004, all her goats and sheep received hay and barley every day and also were taken out to graze in the *kyzk kyzk* group, except for 10 days when the weather was too cold. She said that the winter of 2001 was very cold and her animals had to be given supplementary feed all winter. She never sold any of her animals which were solely for home consumption. By October 2005, Sveta had taken up fishing and moved with the Chu river as the flow went up and down, taking her animals with her.

The headman (*akim*) in one of the study villages commented as follows:

"From ancient times we know that livestock which move put on weight quickly. Next year, we plan to have no animals around the village. The main problem hampering this plan is that though people know it's better to move out, they need help. We wanted to save the pastures around the village, by trying to encourage bigger flock owners to move further out as small flock owners could not find enough grazing around the village. Most big flock owners agreed but water is the main problem preventing smaller flock owners from moving out. In the spring time there are distant pastures with water in the springs but these dry up after a month or so and animals have to be moved again. The artesian wells need to be cleaned from stones, and this requires repair men and costs money. Also there are many pastures with deep wells but these need pumps and diesel. After the late 1990s [when the state farm collapsed], pumps and pipes were stolen for scrap metal to China" (Plate 6).

Small flock owners cannot migrate and take advantage of the variable grazing ecology offered by the Kazakh rangelands, even though they recall this system from the very recent past under the state farm management. Small holders' scale of operation is constrained to raising as many animals as they can feed over winter and herd around the circum-village ranges (Plate 7). Given the high cost per animal unit of purchasing winter feed, small flock owners cannot easily increase their scale of operation due to the marginal cost of adding each animal to the flock. For large flock owners the grazing is essentially fragmented, as only a fragment of the former seasonal nomadic grazing circuits are used, that are accessible to settlements. As shown in the paper by Alimaev et al., this volume, having to retain animals around the villages has led to degradation of the vegetation and soil.

Plate 6: Broken water point for livestock in desert pastures

Plate 7: Overgrazed pastures around Aidarly village

A few families have accumulated large flocks though various means including purchases from family remittances, inheritance, hard work and fortuitous circumstances such as having a strong patriarchal family head able to retain adult family labour. In these cases, then the large flock can be operated on an expansive geographical scale that incorporates the varied seasonal ranges. For large flock owners the grazing circuits remain unfragmented, since there is as yet little formal rangeland privatisation. Examples follow of some of the large flock owners who moved every season with their animals to different pastures and water points (Boxes 3, 4 and 5).

If we consider the expected returns to these two distinct grazing strategies, small flock owners have to pay more than large flock owners to keep each animal, but they gain more per head of animal owned, in terms of the consumption value from their animals. Larger flock owners are commercially-oriented, interested in the cash returns they gain by selling many male animals to markets each year, at between USD $ 50 to 100 for each sheep. They are also interested in providing their animals with what they believe to be optimal natural grazing, especially in the most difficult foraging season of winter when animals taken to the remote grazing grounds in the sands can browse on nutritious seeds and standing hay. Around the villages there is virtually no grazing left in winter, and animals must depend on purchased fodder such as reeds, often of poor nutritional value.

The large flock owners are explicit about the advantages of seasonal migration, despite the considerable financial costs and inconvenience to family life. Their animals are better fed in winter and their offspring born in spring are bigger and mature faster, compared to village-based animals. This

BOX 3. Shepherding for a cooperative flock

Bahitjan was formerly a shepherd employed by the Aidarly sovkhoz (state farm), then for the cooperative of a few families formerly working for the state farm. By 1999 he had a job guarding the pasture research station, and had 130 private sheep and goats as well as shepherding several hundred sheep for the cooperative.

For the winter of 1999/2000 he took the cooperative's animals together with his own to the Sarytaukum sands north of the village for three months. He said "winter in the sands is warmer. If the animals stayed here around the village they would just consume all the hay".

In spring 2000 he moved all the animals 10 km south of the village to the Artemisia semi-steppe pastures, where there was a well but lacking a motor pump. Further south there were better pastures and barns but these had already been claimed by large flock owners in Aidarly, after the sovkhoz collapsed.

In August after the lambs were weaned, the ewes were moved to another well and pasture 12 km away, as he said "they needed to change pastures as the ewes need to get into good body condition after lactation, and before being mated again in October". But the weaned lambs had to remain for the autumn at the spring pastures, as the cooperative had no fuel to move the shepherds' yurts and belongings. Formerly the Aidarly sovkhoz used to take the shepherds for the summer to graze the animals in the Ala Tau mountains several hundred km to the south, but Bahitjan said now he would not take the animals to the mountains as it was a very long distance and dangerous to stay alone there, as few other shepherds came there after the other state farms collapsed.

The cooperative paid him in the form of 3.5 tonnes of barley straw for winter fodder and he also bought 10 tonnes of Agropyron, (wild wheatgrass) for his animals over the winter of 2000/2001.

In summer 2001, Bahitjan lost his job and home as a guard for the research station and also lost his job shepherding for the cooperative, which had broken up. He had a few goats and cattle and was living in a small temporary shack near the pasture research station. By spring 2004, he had accumulated 30 cattle and about 30 goats, but no sheep. His wife was making butter and cream from the cows' milk and selling to villagers. By autumn 2004 he was planning to move out to one of the ruined former state farm barns 20 km south of the village, as he had 15 cattle and about 50 goats. He was working on getting his lease registered for the pasture land surrounding the barn.

means in turn that the market price is comparatively higher for immature animals from larger flocks. One flock owner in Aidarly, owning a middle level of several hundred sheep, remarked:

"I moved my animals north to the sands for winter as it's warmer in the sands, lots of valleys and there is always some grazing left in the valleys. I did not give fodder to most of my animals as they were able to graze the whole time. I only used half the winter fodder I had prepared. If I compare my animals after winter in the sands, to village animals staying around all year, those are lean while mine are fatter. My lambs do not die from being born weak. The village flocks are all the time on one type of vegetation, and in summer there is not enough grazing around the village, therefore animals are not strong enough to have good offspring. Some village cows even die from weakness".

BOX 4. Large flock owners: four brothers with one in the city

Antyk and three adult brothers had combined their animals into a common flock, which had been migratory under their father's command since 1999 when they became private shepherds after the Ulan Bel *sovkhoz* was dissolved. Their father died in 2003, and three of the brothers shared the responsibilities of managing their flock of a thousand sheep and goats, 40 cattle, 6 riding horses and 12 camels. Another brother worked in the main city of Almaty as head of a tax committee, and according to local sources the family therefore had a lot of money to invest in their livestock operations.

The flocks' annual movement cycle in 2003 and 2004 was a foreshortened version of that followed by the former *sovkhoz*, which was itself a truncated version of the long distance movements of the nomads prior to collectivisation in the 1930s. Antyk and his family spent the winter in the Moinkum desert sands south of the Chu river, moving in spring to the flood plain of the Chu, which spreads out when swollen with water from the spring snow melt from its source in Kyrgyzstan's mountains. In June they went north to the Betpak Dalla plains, but only reaching a well 70 km from the Chu river and staying there one month and then moving again to fresh pastures at another well 20 km distant. They returned south to the Chu river in October, for mating the animals, and then moved back further south to their winter pastures. They had bought reed hay and barley for the winter, but the sheep and goats grazed all winter in the sands, and only sick and weak animals were given supplementary fodder, which was mainly given to the riding horses and the cattle. By spring they had one third of the hay remaining, but all the barley had been consumed.

The flock owners, their wives and children took turns living in a large wagon, pulled by a tractor, following their flocks (Plate 8). They had three hired shepherds in winter, and two for the rest of the year, who lived in another wagon. The hired shepherds were itinerant homeless men, who often just worked for one winter and moved on. The shepherds received one lamb for each month worked, in addition to their food and lodging in the wagon. In winter they all lived in a small house the flock owners had bought in the desert pastures, left over from the state farm infrastructure, and where they had rented 500 ha of pasture land around a well, which cannot legally be privatised. By 2006 they had 1,200 sheep, 100 goats, 54 cattle, 10 horses and 16 camels.

But even the largest private flock owners do not take advantage of the best grazing areas within their former migratory circuit. By the time of this study, none of the largest flock owners in the livestock migratory circuit of Aidarly still sent their animals to the southern Ala Tau mountain pastures in summer. This had been the traditional nomadic pattern in the pre-Soviet period and was re-adopted in the 1960s by the Aidarly state farm (Alimaev, 2003; Behnke, 2003).

In the Ulan Bel study site, during the pre-collective and state farm periods, animals would be taken far to the north of the Chu river over summer, more than 250 km, to access the best grazing lands called Sary Arka ("yellow band"). These are the steppe grazing lands with higher precipitation and more grasses, stretching for hundreds of km in middle Kazakhstan. Older shepherding families spoke nostalgically in praise of this summer grazing

Plate 8: Spring camp of large-scale pastoralists 50 km from Ulan Bel on the Chu River flood plain

BOX 5. Large flock owners: sons of stoppe generals

Aidar, his older and younger brothers, and their 80 year old mother, owned in 2001 a total of 400 sheep, 200 goats, 30 camels and some cattle and riding horses. According to local sources, their mother and father had been excellent shepherds in the Ulan Bel *sovkhoz*, received a lot of citations and medals, and were known as "steppe generals". He was first interviewed in August 2001 in the Betpak Dalla plains where he was with the family's livestock, camping in two yurts together with his wife, their young children, one brother, a nephew and a daughter-in-law. He said then that the length of time they would stay at this summer pasture depended on the pastures, and his older brothers knew when and where they would move next. At that time, only 3 other families from Ulan Bel moved up north for the summer grazing. The flocks' annual cycle was as follows:

In spring in late March they moved from their winter house and pastures in the sands up north to near Ulan Bel on the Chu river, for lambing. The sons of five relatives' families in the village came and helped with these tasks. According to Aidar, their spring pasture area is suitable for lambing as it is flatter than the winter pastures further south in the hilly sand dunes where ewes and their new lambs cannot easily be seen. However, he noted that they do not use the spring pastures in winter because they are flat and the there is no protection for animals from the cold winds, so then animals would have to stay more in the barns and get supplementary feeding, which is expensive. They moved again to the flood plains north of the main Chu river in April, for shearing and tick treatments.

(Continued)

BOX 5. (Continued)

They left the Chu river valley in May, going north to Betpak Dalla, explaining that flocks are only moved north when it gets warmer, as the lambs can die if it gets too cold. Aidar also said "Movement from the Chu is determined by the appearance of biting flies which lay their eggs in the reeds and emerge at the end of May. Then we will move further north 100 km to an artesian well. It takes 10 days to reach this well, moving slowly with the wagons" (Plate 9). Over the summer they changed pastures and moved up to 200 km northwards from the river, as he said that it's better to go further north as the ground water in the wells becomes less salty.

One summer, in 2003, they only reached to 50 km north of the river, but at that area there were no "white grasses" (Kazakh: *ak shuup*), four species Aidar named in which the top part (seed head) is whitish/grey and he noted that these grasses are very nutritious for livestock. The white grasses only start to occur 120 km north of the river, beyond Betpak Dalla in the summer grazing area called Sary Arka. In 2004, he commented that "There is a lot of competition now to stay around the wells in Betpak Dalla and the pasture is decreasing", as more large flock owners from Ulan Bel were becoming migratory.

In October they moved back southwards to the spring pasture area, where they stayed until November, then moving 45 km further south to their winter house in the Moinkum sand dunes. For winter of 2003/2004 one of Aidar's brothers who lived in Ulan Bel village had prepared 10 tonnes of reed hay; this brother owned all the hay making mechanised equipment and sold hay to other villagers. Aidar's brother also bought 3 tonnes of barley and wheat bran from the city, using his own vehicle. All this fodder was given only to weak and sick sheep and goats, and to riding horses, but not to cattle or camels. The animals grazed outside all winter in the sands.

In the winter of 2005, Aidar quarrelled with his brother and they split the flock, Aidar and his family moved with their yurt to nearby another large owner from Ulan Bel. This other owner had a truck, which Aidar did not own, and they planned to cooperate.

area: One old lady in 1999 told how she and her family had taken the livestock and their yurts last time to Sary Arka in 1998, but not in 1999 "because of difficulties with transport. We moved to Sary Arka in summer because the pasture there is in a mountainous area, with good rivers and it's like a resort". Another old lady who travelled with her sons and their livestock in 1999 to Sary Arka remarked "Sary Arka is good in summer because it has sweet water from rivers, tall grasses up to the chest, it is not so hot as around here [Moinkum desert] and if you compare the body condition of sheep grazed there in summer it is much better than those grazed around here." In the case of Aidar from Ulan Bel, the son of the "steppe generals", we see that every year he sought to move his family's livestock as far north as possible, to access the sweeter water and steppe grasses, but he lacked motorised transport, and instead had to settle for the more accessible but less optimal pastures in Betpak Dalla for summer.

Plate 9: Summer camp of large-scale pastoralists north of Ulan Bel

There is an intermediate grazing strategy between the sedentary village-based and the seasonally mobile flocks. That is the barn-based grazing system, first adopted by those with larger flocks soon after the end of state farms and government support to livestock production in the mid 1990s. The barns had been built for the state farm flocks, always next to wells in the distant pastures at least 20 km and sometimes much further from the villages, in the case of Ulan Bel former state farm. The barns were used for wintering livestock, or for lambing, shearing, or veterinary treatments. There was often a shepherd's winter house (*zimovka*) built next to the barn. By the late 1990s, these barns had been abandoned after the state farms collapsed and the legal ownership was unclear (Behnke, 2003). Wealthier villagers who had acquired large private flocks at the breakup of the state farms soon took a keen interest in these barns and the wells adjacent to them. The barns presented an opportunity for larger flock owners to provide their animals with better grazing by removing them from the communal pastures around the villages which were rapidly becoming over-grazed as most livestock ceased to migrate seasonally. Such barn owners do not have thousands of livestock heads, which would necessitate seasonal movements as this number could not be kept all year in one location. Flock owners who moved their animals out to a barn had a middle level of livestock wealth with several hundred head of sheep and perhaps 20 or 30 cattle. Sometimes the barn flocks were managed by absentee owners who remained in the village, relying on their sons, nephews, younger brothers and hired shepherds to take care of the animals. Following are some examples (Boxes 6 and 7).

BOX 6. Moving out to a barn: father and son

Janisbek and his grown up son Kazbek had moved out in 2003 to a barn some 20 km south of Aidarly. He had registered 200 ha of pasture land around the barn, under a lease from the district administration. The barn was in the low hills covered with short *Artemisia* and grasses. The barn had formerly belonged to the Aidarly state farm, when it was used for lambing in spring and shearing in autumn. There was a partially destroyed shepherd's dwelling beside the barns, and an artesian well with salty water that was acceptable to livestock. In 2004 he had 270 sheep, 40 goats, a couple of riding horses and a truck.

For the winter of 2003/2004 he bought several tonnes of *Agropyron* hay and barley which was mainly fed to the riding horses; Janisbek said they only fed the sheep and goats females for a few days after lambing and kidding when they stay in the barn. By spring 2004 he had hay left over. He said they were not planning to move to different pastures, as "it would be nice to go to the winter sands, but I am just a beginner [shepherd] as I was a mechanic before. Also if I leave this place and go to the sands in winter, people would come to the barn and house here and remove building materials. Winter in the sands is warmer and the snow is not deep. There is good grazing there".

His wife and two younger sons stayed in the village where they went to school. His older son staying with him at the barn was not yet married, and Janisbek was worried that his future daughter-in-law would not want to stay out at the barn with his son when he got married. They had hired a shepherd over winter, but this young man left to work on the building sites in Almaty as soon as spring arrived.

In October 2004, another son and a hired shepherd were living at the barn and Janisbek was commuting weekly from Aidarly village, to check up.

In April 2006, there was a lack of water in the well at the barn, and animals had to be taken every two days to a larger well about 10 km distant. Their livestock had increased to 300 sheep, 100 goats, and 16 cattle. Janisbek had built a new house at the barn site.

BOX 7. Living at a barn: a young family

Abei, a young man, and his wife with two pre-school children moved out in 2003 to a small winter house by a barn 30 km south of Ulan Bel village. The barn had been bought for USD $ 500 by Abei's wife's father. This man had previously put his private animals into the care of a migrating flock-owner, but Abei's father-in-law had to contribute to the cost of fuel, and fodder for the riding horse of the shepherd with the migrating flock. So Abei's father-in-law decided instead to move all his animals out to the barn and put his son-in-law Abei and his daughter in charge of managing the flock. Abei and his wife owned 15 sheep and 40 goats, which the wife milked and made butter. They had no riding horse and the two small children helped shepherd the animals on foot. They managed about 500 sheep and goats.

The barn and house had been used before by the Ulan Bel state farm to keep young and very old sheep that could not keep up on migrations. Another family from Ulan Bel village had moved with their animals into a neighbouring shepherd house and barn 1 km distant. The two families quarrelled over use of the one well between them, and Abei's family moved away in 2005 to another barn and house some 20 km distant in the Moinkum sands.

By October 2005 they managed 600 sheep and goats belonging to four families from Ulan Bel, but owned only 10 sheep themselves. Their children were attending school in the village, living with relatives. Over winter 2005/2006, Abei was killed in a motorbike accident and his younger brother took over managing the flock.

Plate 10: Outdoor kitchen for seasonally mobile pastoralists

7. Conclusions

Results from previous DARCA research (Kerven et al., 2003 and 2006) in the SCALE study areas supports the findings from the small SCALE sample: animals that were moved gain more weight over winter, but their owners had much higher management costs per unit. On the other hand, sedentary flocks that lose weight over winter bring higher annual economic returns per head to their owners. This higher annual per head income in sedentary flocks reflects a greater offtake rate for home consumption and sales, due to the poverty of non-moving households who have much smaller flocks. However, non-moving small flocks are viable, able to continue providing their owners with an income but not capable of expanding due to high offtake and lack of capital to migrate to seasonal pastures. Large flocks are moved to seasonal pastures, the animals are heavier and lose less weight in winter, flocks yield cash profits, but there are high real costs to maintaining the scale of movement across the landscape (Plate 10). The question remains whether small flock owners will find a way to send their animals on the seasonal migrations, which improve animal productivity and prevent degradation around the villages.

Acknowledgements

We are grateful to Roy Behnke and Grant Davidson for their comments on drafts of this chapter, and to Cara Kerven for her editorial assistance.

Support for this research was provided by the European Commission Inco-Copernicus RTD Project ICA2-CT-2000–10015 'Desertification and Regeneration: Modeling the Impact of Market Reform on Central Asian Rangeland' (DARCA), Macaulay Institute, Aberdeen, U.K., and under a U.S. National Science Foundation Grant No. DEB-0119618 'Biocomplexity, Spatial Scale and Fragmentation.

References

Alimaev, I. I. 2003, Transhumant ecosystems: Fluctuations in seasonal pasture productivity, in: *Prospects for Pastoralism in Kazakhstan and Turkmenistan: From State Farms to Private Flocks,* C. Kerven, ed., Routledge Curzon, London, pp. 31–51.

Alimaev I. I. and Behnke, R. H., 2007, Ideology, land tenure and livestock mobility in Kazakhstan, in: *Fragmentation in Semi-arid and Arid Landscapes: Consequences for Human and Natural Systems,* K. Galvin, R. Reid, R. Behnke and T. Hobbs, eds., Springer, Dordrecht, pp. 151–178.

Alimaev, I. I., Kerven, C., Torekhanov, A., Behnke, R., Smailov, V.K., Yurchenko, V., Sisatov, Zh., Shanbaev, K., The impact of livestock grazing on soils and vegetation around settlements in southeast Kazakhstan in: *The Socio-Economic Causes and Consequences of Desertification in Central Asia*, R. Behnke, ed. Springer Dordrecht, The Netherlands, pp.81–112.

Asanov, K. A., Shax, B. P., Alimaev, I. I. and Pryanishnikov, S. N. 1992, *Pasture Sector of Kazakhstan*, Ghylym, Almaty (in Russian).

Behnke, R. H. and Scoones, I., 1993, Rethinking range ecology: Implications for rangeland management in Africa, in: *Range Ecology at Disequilibrium: New Models of Natural Variability and Pastoral Adaptation in African Savannas,* R. H. Behnke, I. Scoones and C. Kerven, eds., Overseas Development Institute, London, pp. 1–30.

Behnke, R. H., 2003, Reconfiguring property rights and land use, in: *Prospects for Pastoralism in Kazakhstan and Turkmenistan: From State Farms to Private Flocks,* C. Kerven, ed. Routledge Curzon, London, pp. 75–107.

Bekenov, A., Grachev, I., and Milner-Gulland, E., 1998, The ecology and management of the saiga antelope in Kazakhstan, *Mammal Review* **28**(1): 1–52.

Ellis, J., and Lee, R.-Y., 2003, Collapse of the Kazakstan livestock sector: A catastrophic convergence of ecological degradation, economic transition and climatic change, in: *Prospects for Pastoralism in Kazakhstan and Turkmenistan: From State Farms to Private Flocks,* C. Kerven, ed., Routledge Curzon, London, pp. 52–74.

Galvin, K., Reid, R., Behnke, R. H., and Hobbs, N. T., eds., 2007, *Fragmentation in Semi-arid and Arid Landscapes: Consequences for Human and Natural Systems,* Springer, Dordrecht.

Humphrey C., and Sneath D., ed., 1999, *The End of Nomadism?* Duke University Press, Durham, North Carolina.

Kerven C., ed., 2003, *Prospects for Pastoralism in Kazakhstan and Turkmenistan: From State Farms to Private Flocks,* Routledge Curzon: London.

Kerven, C. Alimaev, I., Behnke R., Davidson G., Franchois L., Malmakov, N., Mathijs, E., Smailov, A., Temirbekov, S., Wright I., 2003, Retraction and expansion of flock mobility in Central Asia: Costs and consequences in: *Proceedings of VII International Rangelands Congress*, Durban South Africa, July 26 – August 1, 2003, also published in: *African Journal of Range and Forage Science 2004,* **21**(3): 91–102.

Kerven, C., Alimaev, I., Behnke, R., Davidson, G., Smailov, A., Temirbekov, S., Wright, I., 2006, Fragmenting pastoral mobility: Changing grazing patterns in post-Soviet Kazakstan, in: *Rangelands of Central Asia: Transformations, Issues and Future Challenges*. D. Bedunah, E. McArthur and M. Fernandez-Gimenez, ed., *Rocky Mountain Research Station Serial*, RMRS-P-39 2006, US Dept. of Agriculture, Fort Collins, Colorado.

Martin, V., 2001, *Law and Custom in the Steppe*, Curzon, London.

Olcott, M. B., 1981, The settlement of the Kazakh nomads, *Nomadic Peoples* **8**: 12–23.

Olcott, M. B., 1995, *The Kazakhs*, Hoover Institution Press, Stanford, California.

UNDP 2007, Sustainable Rangeland Management Kazakhstan. http://www.undp.kz/projects/start.html?type = internet

World Bank 2004, Forest Protection and Reforestation Project, Kazakhstan (in Kyzl Orda rangelands) http://web.worldbank.org/external/projects/main?pagePK = 64312881&piPK = 64302848&theSitePK = 40941&Projectid = P078301

World Bank 2003, Drylands Management GEF project Kazakhstan http://web.worldbank.org/external/projects/main?pagePK = 64312881&piPK = 64302848&theSitePK = 40941&Projectid = P071525

Zhambakin, Z. A., 1995, *Pastures of Kazakhstan*, Kainar, Almaty (in Russian).

CHAPTER 7

HUMAN AND NATURAL FACTORS THAT INFLUENCE LIVESTOCK DISTRIBUTIONS AND RANGELAND DESERTIFICATION IN TURKMENISTAN

LIVESTOCK AND DESERTIFICATION IN TURKMENISTAN

ROY BEHNKE[1]*, GRANT DAVIDSON[1], ABDUL JABBAR[2] AND MICHAEL COUGHENOUR[3]

[1] *Macaulay Institute, Craigiebuckler, Aberdeen AB15 8QH, U.K.*
[2] *Mik.-10, Oguz Han Str., Proezd-4 D-11, Korp-2, Kv-28, Ashgabat, Turkmenistan*
[3] *Natural Resource Ecology Laboratory, Colorado State University, Fort Collins, CO 80523-1499, U.S.A.*

Abstract: The ideal free distribution is a biological model that explains the abundance of predators relative to their prey. This analysis reapplies this theory to examine the distribution of domestic livestock relative to the availability of water and forage along a 150 km. transect in the Karakum Desert of Turkmenistan. In this arid environment, the location, quality and quantity of stock water are as important as forage in determining stock movements. Livestock also appear to shift seasonally between preferring high quality or high volumes of forage. This analysis is part of a wider effort to understand why herd owners maintain animal numbers and distributions that promote or retard desertification.

Keywords: Desertification, Turkmenistan, grazing systems, ideal free distribution

1. Introduction

Some aspects of rangeland desertification are better understood than others. The mechanical and biological processes whereby livestock alter rangeland vegetation – by eating, walking or defecating on it – are reasonably well documented. Innovations in rangeland assessment that combine ground-based and satellite

*To whom correspondence should be addressed. Roy Behnke, Macaulay Institute, Cragiebuckler, Aberdeen, AB15 8QH, Aberdeen, U.K; e-mail: roy_behnke@msn.com

R. Behnke (ed.), The Socio-Economic Causes and Consequences of Desertification in Central Asia, 141–168.

technology have also improved the accuracy of environmental monitoring in heterogeneous rangeland landscapes. There is one area, however, where progress has been modest: the attempt to understand the human decisions that cause anthropogenic desertification. In pastoral production systems, we need - at the very least - to understand why pastoralists choose to keep certain numbers and species of domestic animals in particular places for different periods of time.

The kinds of data needed to understand these decisions are as diverse as the constraints and incentives that influence pastoral behavior. In this study we restricted our investigation to five fundamentally different sets of factors that were likely to influence Turkmen herd and flock owners: the availability of forage, drinking water for stock, livestock performance, the economics of alternative husbandry practices, and institutional restrictions on herders' free will. This paper focuses on questions of water and forage availability and livestock performance.

2. Field site and methods

Agro-ecological and socio-economic research was carried out in the pastoral portion of the District (*etrap*) of Goktepe, in Ahal Province (*wiliyat*). Work included surveys of rangeland vegetation and livestock performance, a livestock census and survey of husbandry practices, in-depth interviews with shepherds, farm managers and district-level officials, and the analysis of statistical data available from state organizations and remote-sensed data from satellite sources.

The study area consisted of the pastures in the northern two-thirds of Goktepe District. Goktepe town, the administrative centre of the district, lies about 50 km west of the national capital of Ashgabat on paved roads along the Karakum Canal. The pastures belonging to the district stretch about 150 km to the north of the canal into the Karakum desert. At the time of the study, eleven farmer associations with their headquarters and main settlements along the canal held northern pastures.[1] All of these associations were engaged in irrigated agriculture, but also held state-owned sheep and camels under the care of shepherds permanently resident in the pasture areas. Pastures and settlements north of the canal were accessible only by unpaved desert tracks.

A standard questionnaire was used to collect information on herd composition and size, herd movement patterns over the previous year and the use of fodder. Ninety-two interviews were conducted with randomly selected state and private shepherds along a north-south transect that began at the edge of the settled zone and ran north to the boundary of the district. Rainfall was higher in the southern than in the northern desert pastures - 140 mm per annum in the south versus 110 mm in the north. Groundwater

[1] The farmer associations or *dihan birlishik* of independent Turkmenistan are collective farms that have replaced the Soviet state (*sovkhoz*) and collective (*kolkhoz*) farms. The organization of farmer associations is described in Behnke *et al.* 2005.

was also more plentiful in the south, as waste water from crop irrigation along the Karakum canal was channeled into marshes and lakes in the desert (Plate 3a). The transect included 20 settlements ranging in size from a single family to just under 40 families. None of these communities were engaged in crop agriculture. The pastures covered by the transect supported approximately 21,000 head of sheep and goats, 2,000 camels and, in the most southern settlement on the fringes of the cultivated zone, 130 cattle. Intensive open-ended interviews were held with the managers of collective farms, district-level government staff and shepherds between 1999 and 2004. Plates 1–4 depict landscapes and living conditions in the study area.

3. The dependent variables

Figures 1 and 2 summarize the phenomena we seek to explain, the spatial distribution of livestock in 2002–3. Figure 1 '*The wells of Goktepe, Turkmenistan: grazing areas around settlements*' presents stocking rates for settlements along the Goktepe transect in 2002–03. Two communities - Duwrunli and Hayyr - are geographically isolated and did not share their pastures with neighboring settlements. Otherwise, all communities in the sample lay within 10 km of adjoining settlements - the maximum distance that sheep routinely travel from their base when grazing - and shared their pastures with flocks from neighboring settlements. There were four relatively discrete grazing clusters, two in the sand desert at the northern end of the transect and two in the sand-clay desert in the south.

Grazing intensity in these clusters has been calculated in two ways. One value, total resident stock units per km^2, is based on the total number of livestock that are both owned by the families living in a grazing area and that return to the grazing area for at least part of the year, divided by the total grazing area in km^2. All species of domestic livestock - cattle, sheep, goats and camels - have been converted into stock units equivalent to one sheep.[2] The second measure of grazing intensity, stocking rate expressed in stock units/km^2 of grazing area, refers to the actual period of time livestock spent in a grazing area, 12 months of grazing time per stock unit being equivalent to one resident sheep. In two grazing areas - Durunli and the northern sand desert - the stocking rate in 2002–03 was identical to resident stock numbers because the livestock owned by these communities spent the entire year grazing in the vicinity of their home settlements. For all other grazing

[2] Sheep in the study area were heavier than goats, but we did not consistently obtain information on the proportion of sheep and goats in mixed flocks. For purposes of calculating livestock biomass, we have therefore assumed that sheep and goat weights are equivalent. Based on our own weighing of small stock and a review of the literature on Turkmen camels, we have assumed that one camel equals 4.6 sheep.

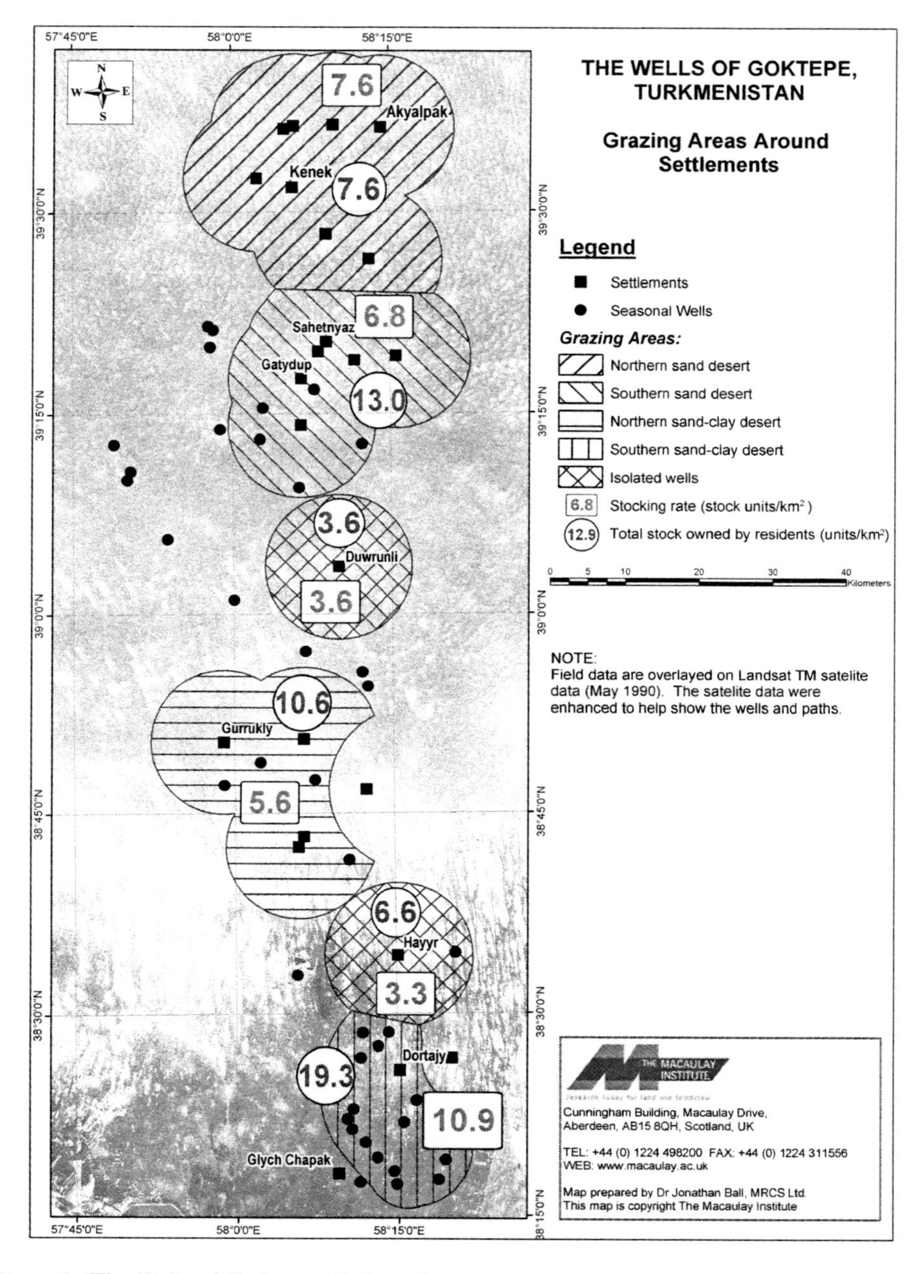

Figure 1. The Wells of Goktepe, Turkmenistan

areas along the transect, stocking rates were lower than resident stock units because livestock moved seasonally away from the settlements to pastures outside a 10 km radius of their home village.

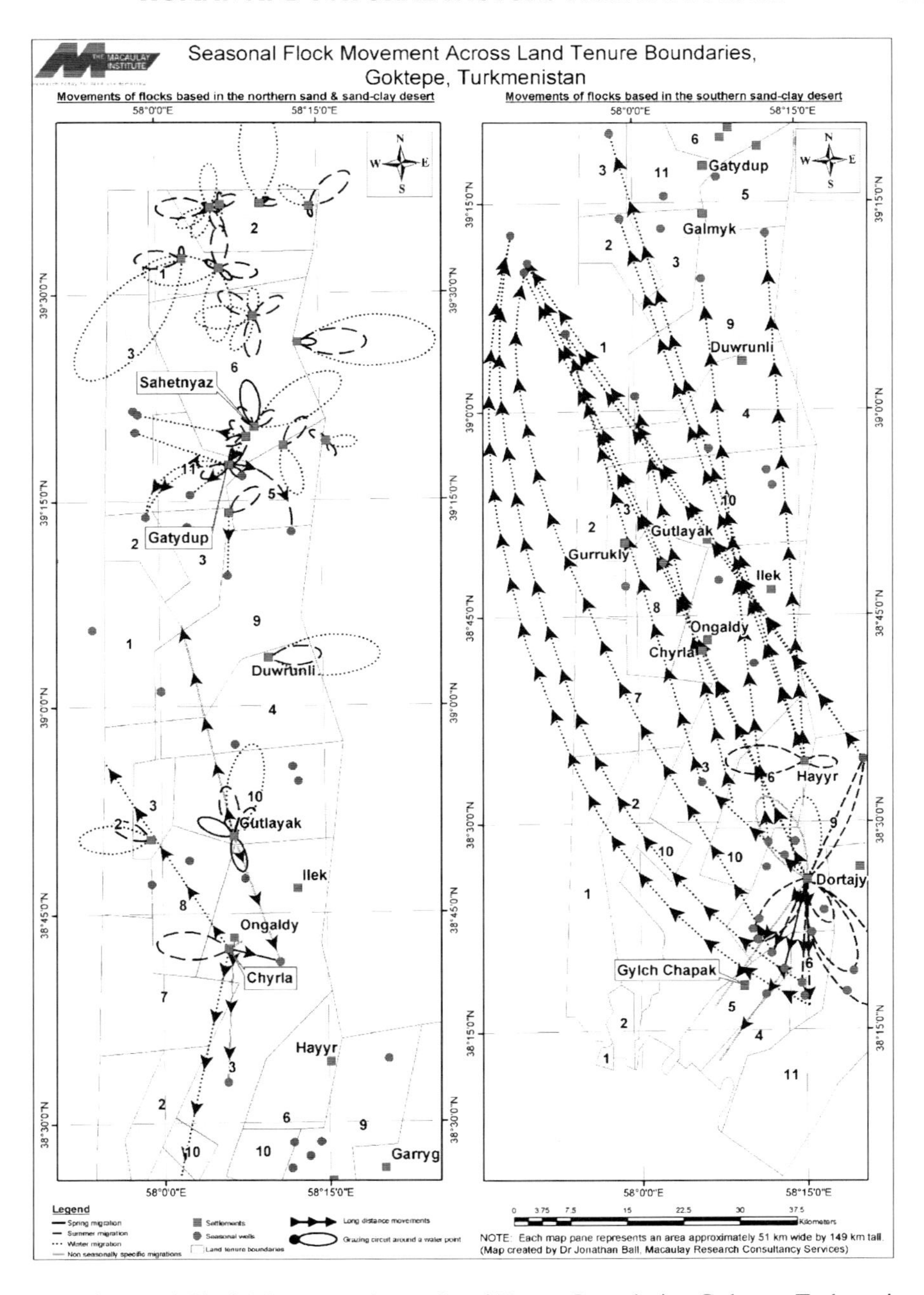

Figure 2. Seasonal Flock Movement Across Land Tenure Boundaries, Goktepe, Turkmenistan

Figure 2, '*Seasonal flock movement across land tenure boundaries, Goktepe, Turkmenistan,*' illustrates the pattern of sheep and goat flock movement in the study area. Elliptical pathways indicate flocks that reside permanently in one settlement. Lines and arrows indicate the direction of seasonal movements.

a

b

Plate 1a: A settlement in the southern desert. Photograph by Grant Davidson
Plate 1b: A settlement in the northern desert. Note the mobile sand dunes in this and the previous photograph. Photograph by Grant Davidson

a

b

Plate 2a: Natural desertification – a solonchak depression or *dupiz*. Photograph by Grant Davidson

Plate 2b: Natural desertification – a solonchak depression or *dupiz*. Photograph by Abdul Jabbar

a

b

Plate 3a: The discharge of irrigation waste water into the desert following a rainstorm. Photograph by Cara Kerven
Plate 3b: Sheep grazing in the northern desert; the woody vegetation is *Haloxylon persicum.* Photograph by Grant Davidson

Plate 4: Shepherd living in a settlement preparing to set off and inspect his flock. Photograph by Cara Kerven

Seasonal movement complicates the task of explaining livestock distributions. In addition to explaining why livestock congregate in certain localities (Figure 1), we must now address several further questions:

- Why do flocks leave certain settlements but not others,
- Why do some flocks leave a settlement while other flocks from the same settlement stay behind?
- Or why do many flocks move in a similar direction at the same time?

Figures 1 and 2 depict the movement and distribution of sheep and goat flocks. Camels, which are an important component of herds, usually do not move seasonally and, in the interests of visual simplicity, their movements are not included in Figure 2. However, the estimates of total resident livestock and stocking rates in Figure 1 include camels, sheep and goats, the numbers of all three species expressed in a standard stock unit. The contribution of camels to total herd biomass varies by settlement and region, with camel numbers generally increasing in the more northerly settlements. As well as the spatial distribution of total livestock biomass, these shifts in herd composition between small ruminants and camels also require explanation.

In sum, the following analysis seeks to explain:

- the spatial distribution of livestock
- their seasonal movements
- the variable species composition of herds.

4. A theory of animal distributions

There is no generally accepted theoretical model to explain the distribution and seasonal movement of free-ranging domesticated livestock. For this study we have borrowed and modified a model developed by ecologists over the last three decades to explain the distribution of predators relative to the abundance of their prey - the ideal free distribution (Fretwell and Lucas, 1970).

Imagine a pond (or create an experimental one) containing many hungry goldfish. Ask people to position themselves around the pond and throw bread into it, one piece at a time so that the bread is consumed almost as soon as it hits the water. Ideal free distribution predicts that the person who throws the most bread will attract the most fish. More precisely, it predicts that if one person throws 20% or 10% or 70% of the bread they will attract, respectively, 20% or 10% or 70% of the fish in the pool (Kennedy and Gray, 1993; Milinski, 1994: 163; Veeranagoudar et al., 2004). This process - termed input matching - is one of the main predictions of 'classical' ideal free distribution theory: the proportion of foragers in a patch matches the proportion of resources in that patch (Ward et al., 2000).

The second prediction of ideal free theory was the 'equal intake' hypothesis, which stated that foragers would eventually distribute themselves so that each obtained the same food intake, despite variable levels of food availability at different sites (Ward et al., 2000; Schaack and Chapman, 2004; Lin and Batzli, 2004). Equal intake rates arise because food abundance in a particular patch and competition for food should, in theory, balance out. If the amount of food at one site is higher than in others, then additional consumers will move into that patch. Movement into the attractive patch will continue until increasing levels of competition between consumers remove the original discrepancy in feed abundance. At this point the incentive to move between patches is gone, the feeding rate is uniform, and the animals have achieved a stable or equilibrium distribution. Implied in this argument is the sequence of patch occupation and resource depletion graphed in Figure 3.

The limitations of early ideal free theory can be ascribed to the convenient but unrealistic assumptions that underpinned it. In the original theory, animals were assumed to be 'ideal' or omniscient in their knowledge of resource distributions, and to be 'free' to travel - quickly and at low cost - to those resources. For simplicity, it was also assumed that these animals were of equal competitive ability and that their foraging behavior was not influenced by predation. These restrictive conditions were always unlikely to be fulfilled under many natural conditions. The lifting of modeling restrictions has, over the decades, diminished the elegance of the initial theory but improved its ability to account for the distribution of natural animal populations (Bernstein et al., 1991; Ward et al., 2000; Winterrowd and Devenport, 2004; Boinski et al., 2003). Strict input matching is now probably best viewed as

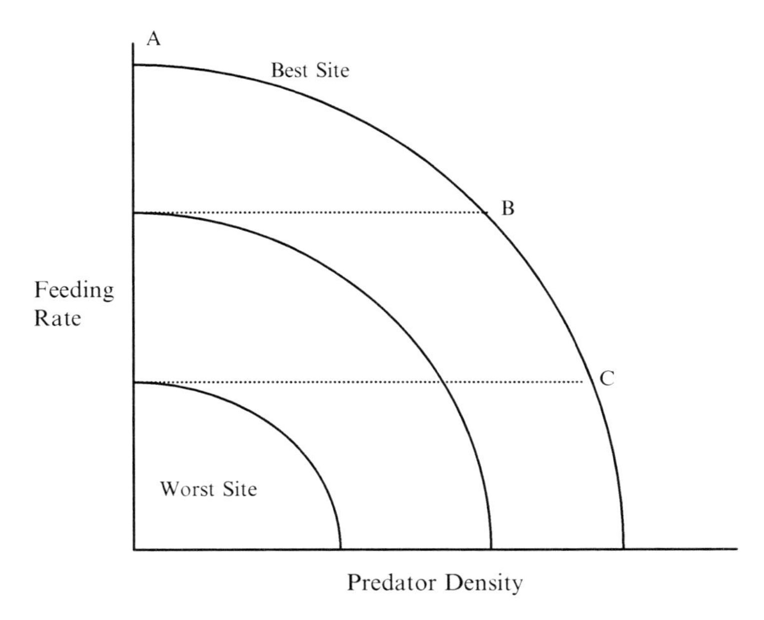

Figure 3. Relationship Between Feedingy Rate and Predator Density. When the number of consumers is low, they should feed only in the best site because feeding rate is highest there (A). As consumer numbers increase, competition will reduce the feeding rate in the best site until it is equal to that in the intermediate site with no competition (B). Consumers should then feed at both sites because the feeding rates are the same. If more consumers arrive, then the feeding rate of both sites will be reduced to that of the poorest site with no competition (C), and they should then feed in all three sites (Sutherland, 1983: 823)

the limiting case of a range of more complex but also more realistic free distribution models (Farnsworth and Beecham, 1997).

While they often were not designed to test free distribution theory, field studies describe ungulate distributions at the landscape scale that are broadly consistent with the matching and equal food intake hypothesizes. This generalization applies to domestic sheep and cattle, mule deer, wapiti, feral horses, North American bison, eastern grey kangaroos and wallaroos, caribou and musk ox (Senft et al., 1987; Wacher et al., 1993; Coppock et al., 1983; Duncan, 1983; Fryxell and Sinclair, 1988; Hanley, 1984; Hunter, 1962). These studies confirm that ungulates track nutritional fluctuations, but a linear one-to-one matching of animal concentrations to feed availability-simple matching - occurs only in restricted circumstances. Complications arise because ungulates are subject to foraging costs and risks that counterbalance nutritional incentives. For instance,

- Ungulates typically lack ideal or perfect knowledge of their nutritional environment.
- They are not free to move without cost within that environment.

- They are subject to predation and disease.
- Individual fitness varies intraspecifically.

These results are consistent with current free distribution theory, which recognizes a variety of matching levels (Huston et al., 1995; Jackson et al., 2004; Farnsworth and Beecham, 1997).

The case of Goktepe illustrates three additional complicating factors that are of recurrent importance in free-ranging Asian and African pastoral systems:

- In Goktepe we are dealing with domesticated animals subject to human manipulation. This being the case, free distributions may be compromised both by economic considerations that make movement too expensive or unprofitable to pursue, or by institutional constraints such as land tenure rights that restrict access to natural resources.
- In semi-arid landscapes such as Goktepe, drinking water for livestock may be as scarce and its availability as determinative of stock distributions as the location of feed. These restrictions provide an example of what can be termed 'constrained matching' in which foraging costs impede movement to sites irrespective of their feed value. The matching of animal numbers to feed availability may still occur, but only in accessible areas. Large areas of potential grazing land are excluded from consideration due to the interaction between abiotic factors such as slope or distance from water and the physical and behavioral limitations of domestic livestock (Arnold and Dudzinski, 1978; Pinchak et al., 1991; Senft et al., 1985; Western, 1975).
- Finally, what constitutes 'prey' in a herbivore system is not straightforward. Herbivores frequently choose grazing areas according to forage quality, the volume of edible biomass on offer being less important than its nutritional characteristics or digestibility whenever it is so abundant that the animals can consume only a small part of what is available. But in seasons of feed scarcity - the dry season in the semi-arid tropics or the winter in the northern latitudes - livestock may reverse their normal preferences and seek abundant feed of diminished nutritional quality. These seasonal shifts in food preference - quantity versus quality - mean that herbivores target different kinds of vegetative 'prey' throughout the course of a year, which may be reflected in spatial shifts in their seasonal distribution (Alimaev, 2003; Van Dyne et al., 1980; Blaxter et al., 1961; Breman et al., 1980; Breman and de Wit, 1983; McNaughton, 1988).

5. Results

The question we seek to answer is: To what extent does the location of livestock in Goktepe constitute an ideal free distribution? The following discussion summarizes a portion of the field data pertaining to this question.

5.1. LIVESTOCK PERFORMANCE AND EQUAL INTAKE

The equal feed intake hypothesis is a central plank of ideal free theory. This study did not directly collect information on livestock feed intake. We do, however, have an indirect measure of intake in the form of sheep and goat weights. Holding factors such as breed, sex and age constant, sheep or goats that weigh the same can be presumed to have equal feed intakes net of foraging costs.

To examine this possibility we selected a sample of 20 flocks in two study communities, one at the south end of the transect in the sand-clay desert and the other to the north in the sand desert. The flocks were visited approximately every three months from August 2001 for eighteen months. During each visit live weights were recorded. For flocks of less than 30, all animals were weighed. For those of more than 30, a representative sample was monitored. For mixed flocks of sheep and goats of over 50 animals, the species were chosen roughly in proportion to the species in the whole flock. To ease identification monitored animals were ear-tagged.

A total of 1,353 small ruminants were weighed and divided into three groups: flocks based in the southern sand-clay desert and migratory, those based in the southern sand-clay desert but resident year-round, and those based in the northern sand desert (see Map 1). Despite the differences in location and husbandry practices, there were no significant differences in adult sheep weights (43.3 kg vs. 43.9 kg; s.e.d. 0.45) between flocks based in the southern sand-clay desert and migratory, those based in the sand-clay desert but resident all year round, and those based in the southern sand desert.

These results are consistent with the equal intake hypothesis of ideal free theory.

5.2. INPUT MATCHING

Input matching – the presumption that the distribution of consumers should be proportional to the distribution of resources – is the second central hypothesis of ideal free distribution theory. Figure 4 provides a profile of the relative availability of forage resources along the transect. Twenty-three settlements in the study area are arrayed along the horizontal axis in Figure 4, with southern settlements to the left and northern ones to the right. The vertical axis is calibrated in Normalized Difference Vegetation Index (NDVI) values, a remote-sensed index of greenness and, hence, an indirect indicator of plant biomass. Figure 4 suggests that there are systematic variations in plant biomass from south to north along the study transect.

The three most southern settlements, data points 1–3 on Figure 4, lie at the base of the Kopekdag Mountains just outside the irrigated zone, an

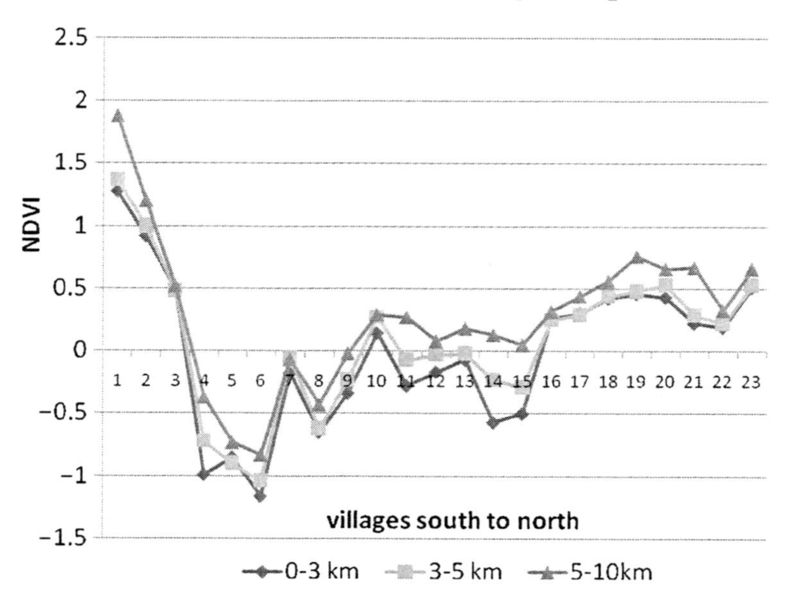

Figure 4. Average NDVI at Variable Distances from Settlements

elevated sand-clay desert area that receives more rainfall than other parts of the transect (Coughenour, 2005). These settlements also have the highest average NDVI. Settlements 4–8 lie in the northern extension of the sand-clay desert, an area of extensive solonchak or hypersaline depressions with very sparse vegetation cover (Plate 2). Settlements 4–6 lie in the midst of these depressions and have the lowest NDVI values of any settlements along the transect. Settlements 7 and 8 lie on the fringes of the concentration of solonchak depressions, and have relatively higher NDVI values, while settlement 9 lies on the locally recognized boundary between the southern sand-clay desert (*ala,* characterized by depressions) and the northern sand desert (*tum*).

NDVI values for settlements in the southern sand desert – settlements 10–15 – are highly variable. Settlements 11, 14 and 15 are situated near three large, adjacent well complexes. Settlements 14 and 15, in particular, are so close together that they constitute virtually a single water source and are so heavily used that there is little vegetation left in the 1–2 km distance between them, a pattern of use that is clear from Landsat images of the region. Settlement 11 is situated around the single largest well field in the study area. All three of these large and closely-clustered settlements have depressed NDVI values relative to smaller and more dispersed settlements in the same region – sites 10, 12 and 13. Moreover, it would also appear that the depressed NDVI values for the large well fields are caused by human use, principally firewood collecting, grazing and vehicle traffic. This conclusion is sustained by the pattern of NDVI values at various distances around these settlements. For the three large settlements, there is a particularly marked difference between NDVI values close to the village at 0–3 km versus those at the edge of the normal grazing radius for village flocks, at 5–10 km. This pattern is consistent with high levels of human disturbance around settlements. By way of contrast, the three smaller and more dispersed settlements in the same region display a much reduced gradient in NDVI values from the centre of the village (0–3 km) to the periphery of the village's grazing catchment, at 5–10 km.

Settlements 17 through 23 lie in the northern sand desert. These water points are relatively small and evenly dispersed. It would also appear that this area may benefit from a slight increase in average rainfall in comparison to areas immediately to its south (Coughenour, 2005). NDVI values for these settlements are consistently high and exceeded only by those at the opposite end of the transect at the base of the Kopetdag mountains. At two sites, 19 and 21, there is a considerable spread between NDVI values close to and distant from the village. The relatively steep gradient in NDVI values around these villages is an indication of probable degradation, since the topography of this area is relatively uniform and the gradients have probably been caused by intense resource depletion close to the villages.

Figure 4 presents a potential resource gradient for the study transect. The preceding commentary suggests that the indexed values in Figure 4 are likely to represent genuine variations in the relative biomass of vegetation around the study communities. There is much that we still do not know, including the dry matter production associated with different NDVI values in different vegetation communities, the amount of the total dry matter that is actually available for grazing, and the nutritional value of this material. Nonetheless, we might hope to see some rough correspondence between NDVI values and stocking rates around the study villages. If this correspondence conformed to the classical or simple version of ideal free distribution theory, then livestock would be clustered (on the analogy of goldfish in an experimental pond) more densely around sites with high NDVI and less densely around settlements with lower NDVI values. Figure 5 examines this possibility.

In Figure 5, village stocking rate is calculated as the total number of livestock using the village grazing area in a year, expressed as sheep equivalents adjusted by the time period actually spent around the village. Figure 5 shows that there is no predictable relationship between NDVI and stocking rates around individual settlements and a regression analysis showed that there was no relationship between these two variables. Figures 1 and 2 in part explain this result: The majority of study settlements are not isolated entities for grazing purposes. Most villages lie close enough together that they share

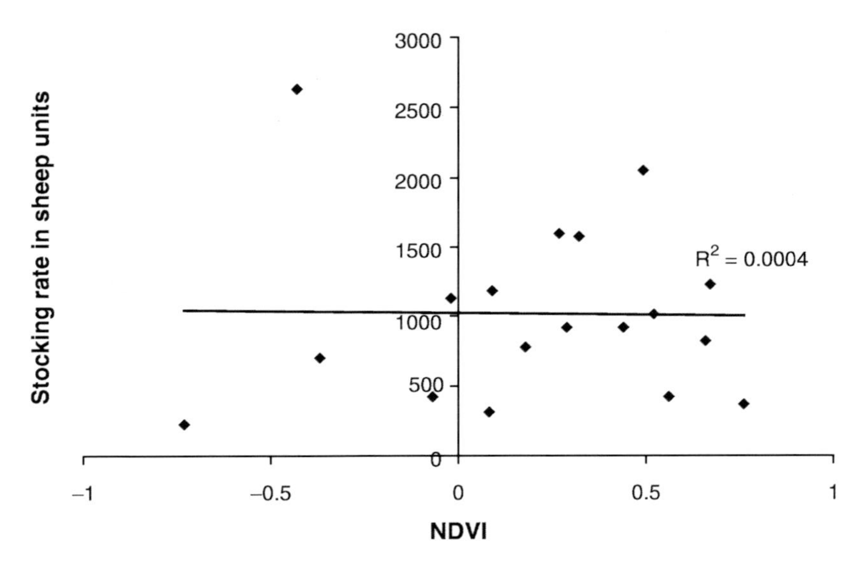

Figure 5. Relationship Between Stocking Rate and NDVI

grazing resources with neighboring communities, effectively becoming part of a cluster of settlements with a common stocking rate. Figure 6 looks at the relationship between NDVI and stocking rates in the areas around grazing clusters, rather than around individual settlements.

Figure 6 illustrates a weak correlation between NDVI and grazing area stocking rate that roughly conforms to ideal free theory, that is, the higher the average NDVI the higher an area's stocking rate. This relationship does not result from the residents in a particular area owning numbers of livestock that reflect the area's NDVI. Rather, residents of areas with high levels of livestock ownership relative to grazing resources relieve excess grazing pressure by seasonally removing their animals from the vicinity of their villages. As a result, stocking rates are both more uniform and parallel changes in NDVI more closely than levels of stock ownership. There are exceptions to this pattern, most obviously in the northern sand-clay desert where heavy concentrations of resident livestock are relieved by out-migration, but at a rate insufficient to bring the stocking rate down to levels comparable to other parts of the transect. In general, however, in areas where levels of stock ownership are high, stocking rates are adjusted downwards by seasonal livestock mobility.

Figure 7 examines the question of local stocking rates from another angle – that of water availability – for eighteen settlements where we know stocking rates and where the livestock also depend primarily on well water.

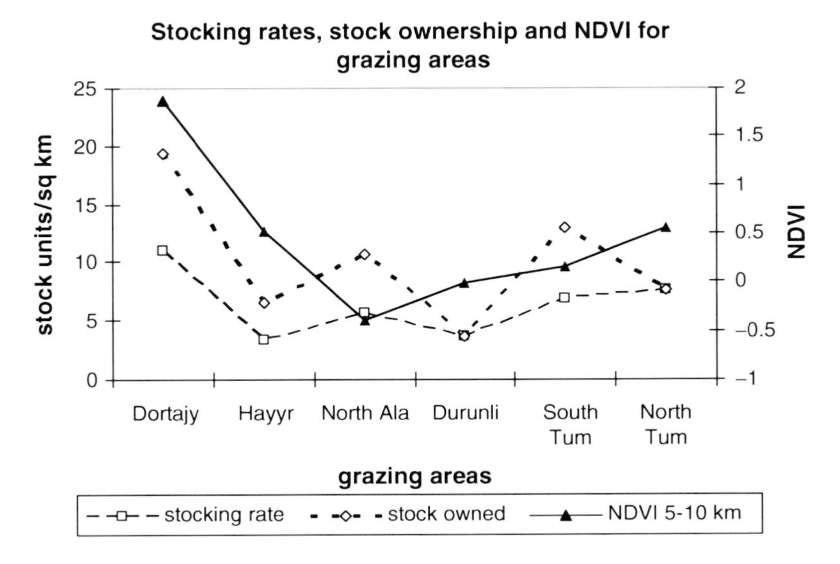

Figure 6. Stocking Rates, Stock Ownership and NDVI for Grazing Areas

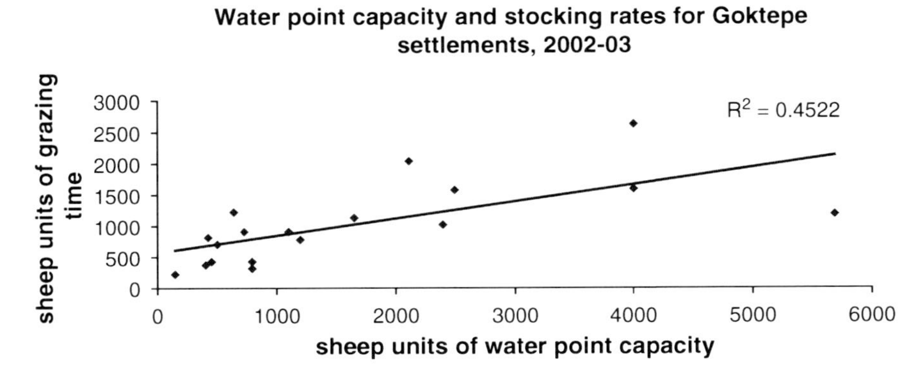

Figure 7. Water Point Capacity and Stocking Rates for Goktepe Settlements, 2002–03

This figure shows that the level of grazing pressure around individual communities reflects the amount of stock water provided by the village's wells and cisterns ($R^2 = 0.4522$). The mean stocking rate around the nine smallest water sources was 670 sheep units, compared to 1,237 sheep units at the nine largest water points, a difference that was statistically significant according to an analysis of variance ($P < 0.05$; s.e.d. 228.8). The notable exception to the overall trend was a pair of settlements located in the southern sand desert within a couple of kilometers of each other that together constituted the single biggest source of well water in the entire study area. If the differential between NDVI close to and far from the village is any indication, the pastures around these villages were also among the most severely degraded in the study area.

In sum, it would appear that the distribution of livestock along the study transect reflects the distribution of two different types of pastoral resources: grazing and water. These two resource types interact in complex ways. At the level of the 'grazing cluster', variations in NDVI are weakly correlated with variations in stocking pressure. At this scale, it would appear that forage availability does influence livestock distribution patterns. It is difficult to go beyond this tentative conclusion because of the imperfect correlation between NDVI and the amount of forage biomass available for consumption by livestock, and because the small sample size makes it difficult to apply statistical tests of significance.

On the other hand, at the scale of the individual desert settlement, stocking rates are controlled primarily not by pasture availability but by water. Apparently, more water attracts more animals up to the point when forage availability becomes a severe limiting constraint. The data presented thus far suggests that water supplies from the very biggest water points may not be used to their full capacity, a density dependent response that merits closer examination in the light of ideal free distribution theory.

Plates 5–9 depict domestic life in pastoral households in Goktepe.

5.3. WATER AVAILABILITY, RESOURCE MATCHING AND DENSITY DEPENDENCE

Figure 8 summarizes data on stock numbers at individual settlements relative to the quantity of stock water available at those settlements. The relationship between livestock density and water availability is presented in two ways. 'Resident stock units' represents the number of livestock units (camels and small ruminants combined) owned by those living at a particular settlement. 'Stocking rate' reflects the period of time these livestock units spent in the vicinity of a well/settlement. In this figure stocking rate is expressed in sheep units of grazing per sheep unit of water point capacity, a measure of the intensity of water use at a particular location.[3]

[3] Turkmen pastoralists express well output in terms of the number of sheep that can be watered continuously before it becomes necessary to leave a well and let it recharge. The process of alternately using and resting wells occurs repeatedly in the dry season, and there is close agreement on well capacity among users. These local estimates are the basis for the water point capacity figures in this analysis.

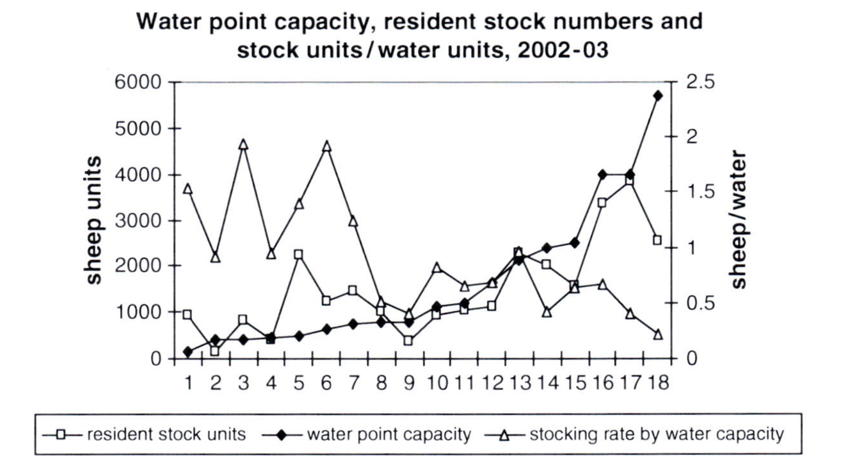

Figure 8. Water Point Capacity, Resident Stock Numbers and Stock Units, 2002–03

Plate 5: Camel milking. Plates 5–9 illustrate the preparation of the staple components of the pastoral Turkmen diet – milk products, meat, and bread. Migratory shepherds living in temporary shelters are often deprived of these products, which are routinely prepared by women. Photograph by Cara Kerven

Plate 6: The finished product - serving *charl*, fermented camels' milk. Photograph by Cara Kerven

For statistical analysis, the 18 settlements in the sample were divided into two groups - those that had a water point capacity of less than 800 sheep units, and those settlements with a capacity of 800 or more units. At a mean capacity of 2,385 sheep units, the watering capacity of the group of large water point points was significantly greater (P 0.01; s.e.d. 604.7) than that of the group of small wells, which had a capacity of 471 sheep units.

From Figure 8 it is clear that there is a strong tendency for large water points to attract more livestock, so that the absolute number of animals increases with the size of the water point. On the other hand, water from large sources is less intensively used than from small sources, which becomes evident if we look at animal numbers per unit of well capacity. Resident livestock/water point capacity was significantly greater (P 0.05; s.e.d. 0.616) in the small wells (2.56) compared to the large wells (0.80). Likewise, stocking rate/water point capacity was significantly greater (P 0.001; s.e.d. 0.148) in the small wells (1.41) compared to the large wells (0.58).

Plate 7: Preparing soup. Photograph by Cara Kerven

These results conform to the theoretical predictions of the ideal free distribution model. On the one hand, there is resource matching, with more livestock congregating around larger wells. On the other hand, there is also evidence for a density dependent response. The rate of increase in animal numbers per unit of available water declines as water availability increases, settlements become larger, and crowding and forage availability become limiting factors. What is unusual about these results is that multiple resources appear to drive the system at the village level – the availability of stock water in smaller settlements and fodder in larger ones.

5.4. NDVI, FORAGE QUALITY AND SEASONAL MOVEMENTS

The preceding analysis has provided a preliminary explanation for the spatial distribution of livestock as depicted in Figure 1. Two additional phenomena remain to be explained – the seasonal flock movements (as depicted in Figure 2) and the variable species composition of herds in the study area.

Plate 8: Baking bread. Photograph by Cara Kerven

Plate 9: The finished product – bread cooling in the yurt. Photograph by Cara Kerven

Seasonal movement and herd composition influence one another. At least in this part of Turkmenistan, camels are not migratory stock. They have the ability to cover great distances quickly, and use this ability to remain in the same area year-round, ranging widely but developing an attachment to their home range that makes it difficult to move them. Forced out of their accustomed range, they simply wander back home. Mixed flocks of sheep and goats, on the other hand, can cover much less ground in a single day and, hence, can use only a restricted area around their home base or water source. But they develop no attachments to these areas and, in compensation for their relative immobility on a daily basis, can be moved long distances seasonally. The map in Figure 2 depicts these seasonal flock movements.

There are three different kinds of seasonal movements to be explained:

1. Localized movements around a stable base – the 'flower petal' patterns characteristic of grazing patterns around villages at the north end of the transect (Figure 2, panel 1). Informants' accounts of these movements provide the basis for the estimation of grazing catchments around villages, but the details of these local movements were not investigated systematically in this study, which concentrated on livestock distributions at the landscape scale.

2. East-west movement within the southern sand desert (*tum*) from permanent village settlements to seasonal winter wells (see Figure 2, panel 1). For flocks in the study area in 2002–3, there were six different pairs of linked village-winter well combinations for east-west movements of this kind. These movements occurred within a single reasonably uniform vegetation community. Comparing NDVI values at the sending and receiving ends of these movements, this kind of movement produced a mean increase in NDVI value of 0.73.

3. Northern movements in winter by sheep flocks from the southern sand-clay desert (*ala*) to the southern sand desert (*tum*) (see Figure 2 panel 2). These movements take place between distinct vegetation communities with different forage characteristics. In 2002–3, there were eleven pairs of linked geographical points of origin-destination for south-north movements of this type, and these movements produced a mean decline in NDVI value of −1.04.

In sum, seasonal movement within a single vegetation community (type 'b' movement east-west within the *tum*) moves flocks to areas of higher NDVI and, presumably, higher total and edible biomass. In this case the NDVI evidence reaffirms statements by shepherds, who say that they move their flocks in winter away from permanent settlements where forage has been heavily grazed in the summer months and into areas that have abundant forage because they have not been grazed in the preceding half year.

South-north type 'c' movements between vegetation communities cannot, on the other hand, be explained in terms of maximizing total biomass, for these moves produce a mean decline in NDVI of about one unit. The probable explanation for these winter movements is that shepherds are seeking better quality forage or an increase in the amount of forage actually available to their animals, rather than an increase in total biomass:

> The contribution of tall shrubs as browse rises in autumn and winter....Over 60 per cent of edible forage in autumn and winter is contributed by shrubs and dwarf-shrubs. In the cold dry season, the above-ground biomass of ephemerals such as *Carex* is dry and dead, while dominant woody shrubs of *Haloxylon* and *Calligonum* can still support browsing animals. Thus the period of maximum yield for some shrubs does not coincide with their maximum forage utility for livestock...*Haloxylon* reaches its maximum yield in summer, when only 5 per cent of that yield is available, whereas in winter when only half of the annual yield is present, 60 per cent is available to livestock. In contrast, the proportion of available forage for *Carex physodes* is relatively constant throughout the year....These seasonal dynamics formed the basis for seasonal livestock movement in the past. Woody shrubs were more prevalent in the northern desert compared to the southern desert, and animals could be taken there in winter to browse when ephemerals in the central desert were senescent (Khanchaev et al., 2003: 198–99).

The processes identified by Khanchaev merit further investigation and documentation.

5.5. WATER QUALITY AND HERD COMPOSITION

Figure 9 introduces a final variable that influences livestock distributions – water quality. Figure 9 presents data on well salinity at settlements and winter seasonal wells arrayed sequentially from south to north along the transect, left to right along the horizontal axis of the diagram. The figure documents the increasing salinity of water points at northern settlements. Figure 9 also shows that winter wells tend to be more saline than wells at permanent settlements at the same latitude. Both of these results verify shepherds' statements.[4]

---

[4] Until the separation is disturbed by extracting water from a well, fresh water sits as a lens on top of salty water and is periodically recharged by infusions of groundwater and surface runoff from rainfall. The apparent salinity of a well in Goktepe therefore depends in part on how recently it has rained and how recently and how heavily the well has been used prior to sampling. While samples were taken over a short period of time, it was not possible to take multiple samples or to rigorously control for levels of prior use. The values in Figure 9 therefore provide an approximate indication of salinity trends that were obscured by our inability to sample more intensively. For example, the 'sweetest' sample from a winter well (and an anomalous value in Fig. 9) comes from a well that had become too salty to use and had been abandoned for the entire year prior to sampling.

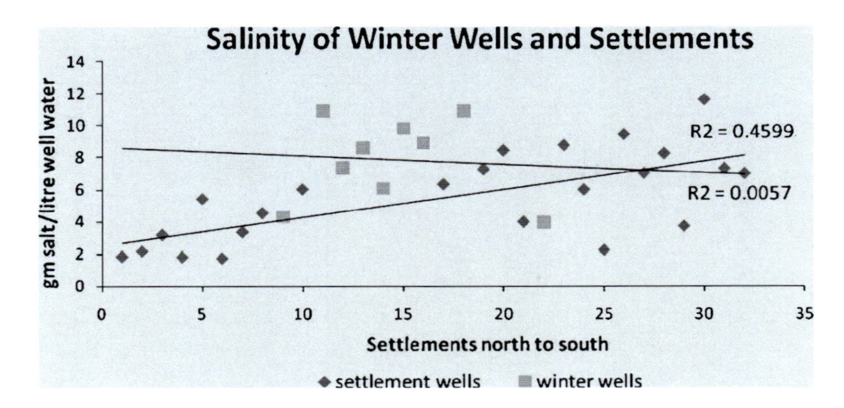

Figure 9. The Salinity of Well Water

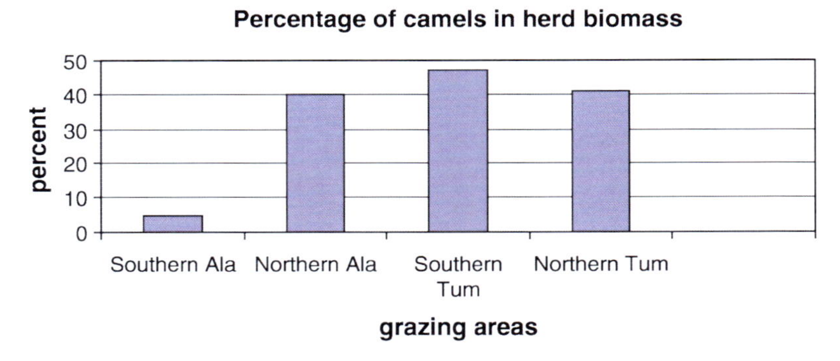

Figure 10. Percentage of Camels in Herd Biomass

According to shepherds, winter wells function as seasonal fodder banks because they are too saline to use year-round. The fodder around these wells becomes available to flocks only in the wet season. At this time of year sheep are physiologically capable of withstanding higher saline levels in their drinking water, their water needs are reduced and it becomes economic to truck fresh water for mixing with the saline well water, and the wells are less saline because of infusions of rain water in this season. Type ‘b’ and ‘c’ migratory movements are therefore caused by a combination of constraints involving both water and forage. Poor water quality around winter wells prevents permanent settlement at these locations, focuses grazing pressure on the areas around villages, and creates a reserve of untouched forage that promotes seasonal exploitation.

The pattern of declining water quality at northern settlements also agrees with shepherds’ statements and helps to explain regional variations in herd composition. From south to north as stock water becomes more saline, camels - which are more salt-tolerant than sheep - become an increasingly prominent component of herd composition (Figure 10).

6. Conclusion

The preceding discussion has presented a preliminary explanation of the spatial distributions, numbers and species composition of pastoral herds - the factors that cause or prevent desertification by livestock in Goktepe. The analysis demonstrates that livestock along the study transect are distributed in a manner consistent with ideal free distribution theory. There is evidence both for a proportional response to resource distributions, with more animals concentrated where there are more resources, and for density dependent processes, as livestock and shepherds respond to over crowding and resource depletion at favorable sites. The patterns of distribution and movement are complex because the system is driven by two variables – forage and water – that are both, in turn, differentiated in terms of their quantity and quality. Because of the overlapping grazing areas between adjacent villages, the operation of forage constraints was most visible at the scale of the grazing area or grazing cluster. The relationship between water availability and stock numbers, on the other hand, was clearly evident at the scale of the individual settlement.

This analysis is far from complete. Household economic data on movement costs explains why large flocks tend to be more mobile than small ones. This data is currently being analyzed. The land tenure system, which explains why certain individuals have priority access to certain resources, to the exclusion of other users, has been described in another publication (Behnke et al., 2005). Integration of these socio-economic factors with the biological variables reviewed here forms the basis for a more comprehensive picture of why and how pastoralists cause or do not cause desertification.

Acknowledgements

Support for this research was provided by the European Commission Inco-Copernicus RTD Project ICA2-CT-2000–10015 'Desertification and Regeneration: Modeling the Impact of Market Reform on Central Asian Rangeland' (DARCA), Macaulay Institute, Aberdeen, U.K., and under U.S. National Science Foundation Grant No. DEB-0119618 'Biocomplexity, Spatial Scale and Fragmentation.'

References

Alimaev, I.I., 2003, Transhumant ecosystems: fluctuations in seasonal pasture productivity, in: C. Kerven, ed., *Prospects for Pastoralism in Kazakstan and Turkmenistan: From State Farms to Private Flocks*, Routledge Curzon, London.

Arnold, G.W., and Dudzinski, M.L., 1978, *Ethology of Free-Ranging Domestic Animals*, Elsevier Scientific Publishing, New York.

Behnke, R., Jabbar, A., Budanov, A., and Davidson, G., 2005, The administration and practice of leasehold pastoralism in Turkmenistan, *Nomadic Peoples,* **9**: 147–170.

Bernstein, C., Kacelnik, A., and Krebs, J.R., 1991, Individual decisions and the distribution of predators in a patchy environment. II. The influence of travel costs and structure of the environment, *Journal of Animal Ecology,* **60**: 205–225.

Blaxter, K.L., Wainman, F. W., and Wilson, R., 1961, The regulation of food intake in sheep, *Journal of Animal Production,* **3**: 51–61.

Boinski, S., Kauffman, L., Westoll, A., Stickler, C.M., Cropp, S., and Ehmke, E., 2003, Are vigilance, risk from avian predators and group size consequences of habitat structure? A comparison of three species of squirrel monkey (*Saimiri oerstedii, S. boliviensis,* and S. *sciureus*), *Behaviour,* **140**: 1421–1467.

Breman, H., Cisse, A.M., Dejiteye, M.A., and Elberse, W. Th., 1980, Pasture dynamics and forage availability in the Sahel, *Israel Journal of Botany,* **28**: 227–251.

Breman, H. and de Wit, C.T., 1983, Rangeland productivity and exploitation in the Sahel, *Science,* **221**(4618): 1341–1347.

Coppock, D.L., Ellis, J.E., Detling, J.K., and Dyer, M.I., 1983, Plant-herbivore interactions in a North American mixed-grass prairie II, Responses of bison to modification of vegetation by prairie dogs, *Oecologia,* **56**: 10–15.

Coughenour, M., 2005, *A Remote Sensing Analysis of Trends in Rangeland Productivity and Condition in Central Asia*, Natural Ecology Resource Laboratory, Colorado St. University.

Duncan, P., 1983, Determinants of the use of habitat by horses in a Mediterranean wetland, *Journal of Animal Ecology,* **52**: 93–109.

Farnsworth, K.D. and Beecham, J.A., 1997, Beyond the Ideal Free Distribution: more general models of predator distribution, *Journal of Theoretical Biology,* **187**: 389–396.

Fretwell, D.D., and, Lucas H.L., 1970, On territorial behaviour and other factors influencing habitat distribution in birds, *Acta Biotheoretica,* **19**: 16–36.

Fryxell, J.M., and Sinclair, A.R.E., 1988, Seasonal migration by white-eared kob in relation to resources, *African Journal of Ecology,* **26**: 17–31.

Hanley, T.A., 1984, Habitat patches and their selection by wapiti and black-tailed deer in a coastal montane coniferous forest, *Journal of Applied Ecology,* **21**: 423–436.

Hunter, R.F., 1962, Hill sheep and their pasture: a study of sheep-grazing in south-east Scotland, *Journal of Ecology,* **50**: 651–680.

Huston, A.I., McNamara, J.M., and Milinski, M., 1995, The distribution of animals between resources: a compromise between equal numbers and equal intake rates, *Animal Behaviour,* **49**: 248–251.

Jackson, A.L., Humphries, S., and Ruxton, G.D., 2004, Resolving the departures of observed results from the Ideal Free Distribution with simple random movements, *Journal of Animal Ecology,* **73**(4): 612–622.

Kennedy, M. and Gray, D.G., 1993, Can ecological theory predict the distribution of foraging animals? A critical analysis of experiments on the Ideal Free Distribution, *Oikos,* **68**: 158–166.

Khanchaev A.H., Kerven, C., and Wright, I.A., 2003, The limits of the land: pasture and water conditions, in: C. Kerven, ed., Prospects for Pastoralism in Kazakhstan and Turkmenistan: From State Farms to Private. Flocks, Routledge Curzon, London.

Lin, Y.K. and Batzli, G.O., 2004, Movement of voles across habitat boundaries: effects of food and cover, *Journal of Mammalogy*, **85**(2): 216–224.

McNaughton, S.J., 1988, Mineral nutrition and spatial concentrations of African ungulates, *Nature,* **334**: 343–345.

Milinski, M., 1994, Ideal Free theory predicts more than only input matching - a critique of Kennedy and Gray's review, *Oikos,* **71**(1): 163–166.

Pinchak, W.E., Smith, M.A., Hart, R.H., and Waggoner, Jr., J.W., 1991, Beef cattle distribution patterns on foothill range, *Journal of Range Management,* **44**: 267–275.

Schaack, S., and Chapman, L.J., 2004, Interdemic variation in the foraging ecology of the African cyprinid, *Barbus neumayeri*, *Environmental Biology of Fishes,* **70**(2): 95–105.

Senft, R.L., Rittenhouse, R.L.R., and Woodmansee, R.G., 1985, Factors influencing patterns of cattle behavior on shortgrass steppe, *Journal of Range Management*, **38**: 82–87.

Senft, R.L., Coughenour, M.B., Bailey, D.W., Rittenhouse, L.R., Sala, O.E., and Swift, D.M., 1987, Large herbivore foraging and ecological hierarchies: landscape ecology can enhance traditional foraging theory, *BioScience*, **37**(11): 789–799.

Sutherland, W.J., 1983, Aggregation and the 'Ideal Free' distribution, *Journal of Animal Ecology*, **52**: 821–828

Van Dyne, G.M., Brockington, N.R., Szocs, Z., Duek J., and Ribic, C.A., 1980, Large herbivore subsystem, pp. 269–537 in *Grasslands, Systems, Analysis and Man, International Biological Programme 19*, Cambridge University Press, London.

Veeranagoudar, D.K., Shanbhag, B.A., and Saidapur, S.K., 2004, Foraging behaviour in tadpoles of the bronze frog Rana temporalis: experimental evidence for the Ideal Free Distribution, *Journal of Biosciences*, **29**(2): 201–207.

Wacher, T.J., Rawlings, P., and Snow, W.F., 1993, Cattle migration and stocking densities in relation to tsetse-trypanosomiasis challenge in the Gambia, *Annals of Tropical Medicine and Parasitology*, **87**(5): 517–524.

Ward, J.F., Austin, R.M., and Macdonald, D.W., 2000, A simulation model of foraging behaviour and the effect of predation risk, *Journal of Animal Ecology*, **69**: 16–30.

Western, D., 1975, Seasonality in water availability and its influence of the structure, dynamics and efficiency of a savannah large mammal community, *East African Wildlife Journal*, **13**: 265–286.

Winterrowd, M.F., and Devenport, L.D., 2004, Balancing variable patch quality with predation risk, *Behavioural Processes*, **67**(1): 39–46.

PART III

CASE STUDIES OF RESOURCE DEGRADATION AND DESERTIFICATION CONTROL

CHAPTER 8

LAND REFORM IN TAJIKISTAN: CONSEQUENCES FOR TENURE SECURITY, AGRICULTURAL PRODUCTIVITY AND LAND MANAGEMENT PRACTICES

CONSEQUENCES OF LAND REFORM IN TAJIKISTAN

SARAH ROBINSON*[1], IAN HIGGINBOTHAM[2], TANYA GUENTHER[3] AND ANDRÉE GERMAIN[4]

[1]*La Cousteille, Saurat, 09400, France*
[2]*Foreign Affairs and International Trade Canada, 125 Sussex Dr., Ottawa, ON, K1A0G2, Canada*
[3]*International Medical Corps, 1313 L St. NW, Washington, DC 20005, U.S.A.*
[4]*Dept. of Epidemiology, University of Ottawa, 451 Smythe Rd., Ottawa, ON, K1H8N5, Canada*

Abstract: This paper examines the impact of land reform on agricultural productivity in Tajikistan. Recent legislation allows farmers to obtain access to heritable land shares for private use, but reform has been geographically uneven. The break-up of state farms has occurred in some areas where agriculture has little to offer but, where high value crops are grown, land reform has hardly begun. In cases where collectivized farming persists and land has not been distributed, productivity remains low and individual households benefit little from farming. Where distribution has occurred some households have prospered, yet many have been left landless or with insecure tenure. Unsustainable use of soils is most likely to occur amongst these groups, the poorest in rural Tajikistan, as they farm the most marginal land and are the least able to access fertiliser. Pastures are still accessed by all households and usually managed as communal property; the legislation favours privatization, but the implementation of this is only just beginning.

Keywords: land tenure; Tajikistan; Central Asia; reform; land degradation; poverty; agricultural systems; pasture management

* To whom correspondence should be addressed. Sarah Robinson, La Cousteille, Saurat, 09400, France; e-mail: sarah.robinson@orange.fr

R. Behnke (ed.), The Socio-Economic Causes and Consequences of Desertification in Central Asia, 171–203.

1. Introduction

The way in which rural households access land greatly influences agricultural productivity, the environmental sustainability of farming and ultimately the living standards of land users. The directions taken by land reform in Central Asian countries will be crucial to changes in these three areas.

In Central Asia during the Soviet period, farming was carried out collectively by salaried workers on state owned land and decisions taken by farm managers were based on plans elaborated by central government. In Tajikistan, since independence in 1991 there has been a process of both privatization and individualization of farming. These two terms have quite different meanings. Many farms which are today registered as private enterprises still retain their collective structures and work much as the former state farms. Others distributed land to households, but although some of these may have permanent user rights over a land share, others rent their land from the collective administration. Lerman (2000) and Lerman *et al.* (2002) have shown empirically that amongst restructuring countries of the CIS and Eastern Europe agricultural performance is directly related to the extent to which farming has been *individualized.* These authors argue that Central Asian countries are amongst the worst performers and they link poor agricultural output to a high persistence of collective structures in which privatization has consisted simply of changing the sign on the door. Macours and Swinnen (2000), show that strong user rights and liberalization are positively related to agricultural labour productivity in post-Soviet economies.

Even where farming is managed at the household level, tenure regimes influence and even determine farming practices. A number of studies have found links between the security of tenure of household farms and soil conservation practices. Mink (1993) and Southgate (1988) mention insecure tenure combined with population pressure and a movement of small farmers to fragile marginal lands as primary reasons why small farmers cultivate their land to exhaustion. Many are unable or unwilling to invest in a resource that they do not own. Neef (2001) found that in Benin soil conservation measures such as tree planting were much more common amongst those with long term use rights whilst the use of mulching and crop rotations was much higher on land that was not subject to conflicts about ownership or use rights. In Niger those with short term land use rights were less likely to use manure than those with longer term rights (Neef, 2001).

The tenure of pastures presents a quite different set of issues from those related to cultivated land. Privatization or leasing by individuals is under debate in many Central Asian countries, while a large body of literature suggests that, under the right conditions, regulated common property regimes (which are currently the norm in Tajikistan) may be the better way forward to avoid exclusion of poorer users whilst avoiding degradation (Ostrom 1990, Wen Jun Li *et al.*

2007, Rohde *et al.* 2006). In light of this debate, we also examine the legislation governing pasture tenure in Tajikistan and the actual patterns of use.

Tajikistan is an interesting country in which to examine the links between land reform and land use. There exist examples of both collective and family farms, and amongst the latter many degrees of security of tenure. We are also able to examine the links between poverty, tenure and land use as data on income is available for many of the household farm types examined in this paper. Tajikistan is also a food-insecure country with high population growth; it has the lowest irrigated land to population ratio in Central Asia[1], while the population is predicted to grow from 6.6 million in 2003 to 7.3 million in 2015 (UNDP, 2005). Migration is one consequence and the International Organization for Migration estimates that 630,000 or 18% of the economically active population left Tajikistan between 2000 and 2003 (Olimova and Bosc, 2003).

There is a lack of field data on changing soil quality in Central Asia, and indeed this is the case for most developing countries. Although the use of soil conservation methods (fertilisers, tree planting, manure application) are extensively documented in many countries, there are few empirical studies that actually link these behaviours to changes in soil quality (soil organic carbon, nitrogen levels, soil structure) in the field. In this paper we look at security of tenure and new patterns of farming and productivity amongst different types of reformed farms and differing income groups and suggest what these might mean for future agricultural sustainability. We present a brief overview of the legislative and institutional frameworks under which farming and livestock grazing are conducted in Tajikistan and describe the large regional differences in the extent to which land reform has taken place. We describe how the differences in behaviour between collective and individual structures and between the winners and losers in the land reform process may lead to poor land management and low productivity amongst specific groups.

2. The legal framework for land reform in Tajikistan

On paper, land reform began quickly after independence in 1991 with over 30 laws and decrees issued pertaining to land since that time. The Land Code of 1996 and its later amendments form the basis of land legislation. The law of May 2002 'On dekhan farms[2]' lays out in detail the framework under which

---

[1] In 2004 Tajikistan had 0.1 ha of arable land per capita (State Statistical Committee 2005). Goletti and Chabot (2000) estimated the availability of arable land per capita to be the lowest in Central Asia, three times lower than in neighbouring Kyrgyzstan.

[2] Dekhan means a farmer in Tajik, but the term dekhan farm can in fact be more closely translated as private farm in English. In this paper we refer to dekhan farms as farming enterprises with permanent land use rights as per the 2002 law. All land legislation cited here is taken from the reference: Land Committee of the Government of Tajikistan (2004).

individual private farming occurs today. Land remains state property but all citizens who are able to work have the right to establish a dekhan farm (*khojagi dekhkoni* in Tajik) with permanent heritable land use rights (Article 6, On dekhan farms, 2002). These dekhan farms may be created on former *sovkhoz* (state farm) or *kolkhoz* (collective farm) land or from other land known as the state fund. There are two levels of documentation required for dekhan farms. The *land certificate* showing the physical size and location of the farm is the document held by the manager or leader of the dekhan farm. Members, who may be from the manager's own household or from other participating households, hold *land share documents*. According to a presidential decree of 31 March 2001 the cost of the land certificate is 63 somoni ($18.5), registration being paid for by the dekhan farm. Costs of land survey and cadastre are paid by the state in theory (Article 10, On dekhan farms, 2002) but in reality the applicant usually pays.

In Tajikistan those wishing to establish their own dekhan farm from state fund land have to apply to the raion (district) government or *Hukumat*. It is the applicant who proposes the number of hectares and number of members (Article 14, On dekhan farms 2002). In the case of former *sovkhoz* and *kolkhoz* lands legislation provides for a mechanism of egalitarian distribution of shares to permanent workers[3]. However there is no real definition of permanent workers in the law, leaving vague who is truly entitled to these shares.

A number of decrees were issued forming a timetable for restructuring of *sovkhoz* and *kolkhoz*. According to the last decree, issued in October 2002, the remaining collective state enterprises, except the seed producing, nursery and stockbreeding farms, were to be restructured by 2005. However restructuring does not just mean the creation of dekhan farms. The Presidential degree 'On the reorganization of agricultural enterprises and organizations' of 1996 specifies that farms may also be reorganized into collective enterprises of various types. In these cases the 'permanent workers' do not actually receive their shares but remain members of a larger unit. If they want to actually receive physical (as opposed to paper) shares then they have to apply to the raion land committee. During the restructuring process the raion land committee is responsible for deciding how much of the former collective land should become state fund land. This land may be available for renting, the establishment of new dekhan farms or other uses at the discretion of local government.

Leasing of land is possible, but the legislation on dekhan farms specifies that dekhan farm land may be leased out only in the case of poor health of

[3] Article 15 of On dekhan farms (2002) and the Presidential decree on reorganization of agricultural enterprises and organizations, 1996, point 21.

the legal land users. The 1996 Land Code (Article 14) states that the maximum length of contract is 20 years, rates are centrally set and contracts may be terminated for non-use or environmental degradation (Article 38).

Although the 1996 land codes prohibited sales of user rights, according to Giovarelli (2004) local governments are able to reallocate dekhan farm land for use by another private party, resulting in high tenure insecurity. According to Article 18 of 'On dekhan farms', land users are obliged to 'promote effective use of the land so as to increase the harvest', and Article 31 specifies that land may be taken back for reasons including degradation of the land. The 1996 land code stated that land may be withdrawn for 'non-rational use' or if is not used for one year (Article 37). Note that in combination, the "non-use" and soil degradation provisions are in direct opposition as, if a farmer allows his/her land to lie fallow in order to improve soil fertility, he/she risks having it taken away.

Apart from land shares for dekhan farms, some additional land was provided to the population in the form of 'presidential land', small plots that were distributed by presidential decrees (in 1995 and 1997) to those members of the population who had smaller kitchen gardens than the national minimum. Kitchen gardens are owned by most rural households in cultivable areas of the former Soviet Union and consist of small plots adjacent to the house. Thus almost 100% of rural Tajiks have access to this type of land. In this paper we will refer to presidential land and kitchen gardens together as 'household plots'. There is a high regional variability in the area of agricultural land available for farming, and private shares given to dekhan farmers in one region may be smaller than household plots in another, however, within any given region, household plots are far smaller than agricultural land shares obtained through the privatisation process and are thus considered separately in this study.

Although pasture is not specifically mentioned in the 2002 law 'On dekhan farms', in the 1996 Land Code it figures amongst the land types available for privatization, and thus the processes followed for pasture allocation are theoretically the same as those for arable land. In reality however, there are many differences in the way in which tenure of pasture and arable land have evolved.

3. Sources of field data on land reform in Tajikistan

Below we discuss the application of land reform and resulting farming patterns in four oblasts[4] of Tajikistan for which data are available. These are listed in Table 1 and shown in Figure 1, and the studies which are the source

[4]Oblast is the Russian word for a region. The Rasht Valley is not part of an oblast but comprises a number of raions (districts) which report directly to Dushanbe and which are known as Raions of Republican Subordination (RRS).

TABLE 1. Sources of data on land reform in Tajikistan

Oblast	Raion	Data source	Sample size (households)
Gorno-Badakhshan Autonomous Oblast (GBAO)	All seven raions plus Khorog town	1. Household survey by MSDSP (2003)	1. 700
		2. Participatory Impact Assessment MSDSP (2004)	2. 208 plus key informant interviews
The Rasht valley (RRS*)	Faizabad, Rogun, Nurabad, Rasht, Tavildara, Tajikabad and Jirgatal	1. Household survey by MSDSP (2004)	1. 867
		2. Participatory Impact Assessment AKF/ MSDSP (2005)	2. 181 plus key informant interviews
Khatlon Oblast (Mountainous)	Shurabod, Mominobod and Khovaling	1. Household survey by MSDSP (2004)	1. 1000
		2. Client Perception Survey AKF (2005)	2. 247 plus key informant interviews
Khatlon Oblast (cotton growing)	Shaartuz, Bokhtar, Kolkhozabad, Pianj, Khabodian	Land Reform in Tajikistan, from the Capital to the Cotton Fields, Action Against Hunger (Porteous 2003).	1000 plus key informant interviews
Sughd Oblast	Isfara	Access to Justice Survey AKF/SDC (2005)	260 plus key informant interviews

*Raions of Republican Subordination

of the data presented in this paper are also listed. Much of the data are from household surveys undertaken by the Mountain Societies Development Support Programme (MSDSP), a locally registered NGO and a project of the Aga Khan Foundation (AKF). The surveys were based on quantitative analysis of data collected through household questionnaires. In addition to these surveys MSDSP undertook qualitative analytical studies. Participatory Impact Assessments (PIAs) looked at factors behind changes in living standards in the Rasht Valley and Gorno-Badakhshan Autonomous Oblast (GBAO). Secondly a Client Perception Survey (CPS) was undertaken to look at access to services by households in mountainous areas of Khatlon Oblast.

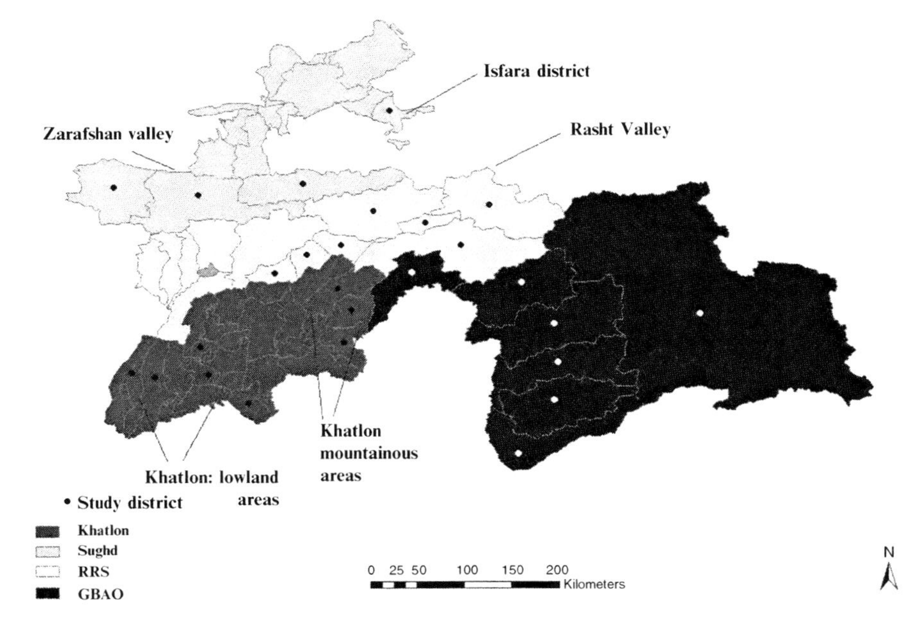

Figure 1. Districts for which field data on arable land reform are available

In the autumn of 2005, AKF Tajikistan in collaboration with the Swiss Office for Development and Cooperation (SDC), designed and implemented the "Access to Justice" baseline survey on behalf of the nascent Isfara Legal Aid Centre in order to identify priority judicial issues facing residents of Isfara City and the surrounding areas. This study contained a large component on land rights. Action Against Hunger conducted a large study of land tenure in cotton growing areas in Khatlon Oblast in 2003 which includes quantitative and qualitative data on the progress of land reform in this region (Porteous, 2003). With the exception of this last study, all the studies listed here were designed and supervised by the authors of this paper.

4. The application of land reform

The data are summarized in figures in Table 2, presenting access to various different land types for each of the regions. The categories of land tenure given in the table are briefly described here.

There are two types of dekhan farm – individual dekhan farms (IDFs) and collective dekhan farms (CDFs). Individual dekhan farms are family-based entities that have a land certificate, and in which each member has

TABLE 2. Access to different types of land by households from six surveys conducted between 2003 and 2005

	GBAO (2003)		Rasht (2004)		Khatlon mountainous (2004)		Khatlon mountainous (2005)	Khatlon cotton (2003)	Sughd (Isfara[***]) (2005)
	MSDSP Household survey		MSDSP Household survey		MSDSP Household survey		Client Perception Survey (CPS)	Action Against Hunger	Access to Justice study
Region	Access	Av size	Access	Av size	Access	Av size	Access	Access	Access
Data source	(% hh[****])	ha	(% hh)	ha	(% hh)	ha	(% hh)	(% hh)	(%hh)
Dekhan farm[**]	72%	0.35	16%	1.21	5%	4.32	29%	3.5%	2.6%
Rented land	16%	0.69	27%	0.65	78%	1.47	48%	6.8%	5%
Presidential land	-	-	17%	0.24	42%	0.32	40%	70%	16.5%
Kitchen garden	84%[*]	0.09	98%	0.15	97%	0.13	91%	99%	64%
Access to land other than household plots	**86%**[*]	-	**41%**	-	**82%**	-	**75%**	**10%**	**5%**

*These figures correspond to about 100% of the population living in areas where land is available. The 14% without land live in Khorog town or Murghab raion where there is no land for distribution.

**Meaning either ownership of land certificate for an individual dekhan farm or land share document in a collective dekhan farm, with access to a *physical* share.

***These results reflect patterns in southern and eastern parts of Isfara Raion, Isfara city dwellers with access to land are negligible.

****Household

a share document. An IDF may include a number of families, but those families enter into the process willingly and decide to apply together for one certificate, usually to save registration costs. Collective dekhan farms correspond geographically to former *sovkhoz* or *kolkhoz* in which former workers have become share holding members. In some regions of the country these members have received physical shares and share documents, and are thus farming as individual households. In others they have received only virtual shares and are actually labourers for what is in reality a collective farm. In both cases the certificate remains with the head of the CDF. There are also some unreformed state farms known by various names, the most common of which is state enterprise, in which workers rent land parcels from the collective. This type of structure persisted until 2005 which was the deadline for reform, but all seed producing and livestock breeding farms remained as state enterprises even after this date. In Table 2 the term dekhan farm refers to cases in which the households in question have either their own IDF or have received physical shares in a CDF. Thus, they are farming as individual households with permanent user rights. Members of CDFs who have not received physical shares and members of unreformed state enterprises either rent land parcels or are landless.

The data indicate that the highest levels of individual access to long term land rights is in Gorno-Badakhshan Autonomous Oblast (GBAO) where almost 100% of households in areas with available land (household plots) in fact have received a private share, with a small percentage of renters. In mountainous Khatlon about 80% of households have gained physical access to some kind of agricultural land apart from kitchen and presidential land (household plots) through either privatization or renting. The proportion of accessed land that is registered as part of a dekhan farm increased greatly between 2004 and 2005. In Rasht the percentage of households with access to land other than household plots is 41%. In cotton growing areas in Khatlon the figure is 10%, in Isfara it is about 2%. In the next sections we discuss how and why these differences have occurred, and subsequently discuss the tenure of pastures.

4.1. CONTRASTING PATTERNS OF REFORM IN FOUR OBLASTS

In Gorno-Badakhshan the local government undertook land reform in the late 1990s jointly with MSDSP. At the time this NGO was providing food aid to a region that was producing only about 15% of its caloric needs and facing a famine situation. It made agricultural inputs available to the population on credit in exchange for co-operation from the local government on land reform. Land was actively distributed to all former *sovkhoz* and *kolkhoz* members with each family receiving land depending on the number of members and the amount of land available for distribution in the former collective of which they were members. This process was over before land reform had begun in the rest of the country, and it was very similar to that which took place in Kyrgyzstan around the same time. It led to an equitable distribution of land in which the individualization of farming was virtually complete (Plate 1). From a legal point of view each former collective or state farm became a collective dekhan farm (CDF) and each land holder a member of this collective. The process of distribution of certificates and share documents is still ongoing but the *de facto* security of tenure is felt to be good by most land holders. Such security is important but in this highly mountainous area average land share parcels are just 0.35 ha, not much bigger than household plots in other parts of the country (Plate 2). Thus although the reform has helped to improve food security, longer term economic effects are likely to be modest.

In the Rasht Valley (Plate 3) land is valuable as much of it is irrigated and, unlike GBAO, the areas available are larger and closer to markets, allowing for the possibility of commercial farming. MSDSP surveys revealed that state or collective farms, in the sense of a centralised management of production undertaken by paid labourers, usually disappeared quite quickly or continued on a small proportion of the land. State enterprises persisted with no formal restructuring for a number of years and households rented land from the farm manager, paying either in kind or cash. By 2004 16% of households were recorded to have individual private farms, which mostly belonged to those who were either able to pay the necessary bribes or circumvent them via contacts. They usually received land which had been part of a *sovkhoz* or *kolkhoz,* and which by law should have been distributed equally to members. This naturally led to a situation of 'first come first served'.

Plate 1: Farmers planting potatoes on land shares in Gorno-Badakhshan Autonomous Oblast. Photograph ©Matthieu Paley/www.paleyphoto.com

Plate 2: Aerial view of Barchidev village, Gorno-Badakhshan Autonomous Oblast. In much of GBAO the area available for irrigated agriculture is tiny whilst rainfall is too low and slopes too steep for rainfed agriculture. Such villages are also prone to natural disasters such as landslides. Photograph © Aga Khan Foundation/Jean-Luc Ray

Plate 3: Winnowing wheat and collecting firewood in the upper Rasht Valley. Photograph © Aga Khan Foundation/Jean-Luc Ray

The majority of the population did not apply at all due to the cost of certification and lack of information – many did not even realise that it was possible to apply for land. As the 2005 deadline for farm restructuring loomed many former collectives restructured formally as CDFs, offering physical shares of an equal size and proper title documentation to all members. However even in communities where this process had occurred many families did not receive land as they could not pay for the share titles, did not have enough labour to farm the land or were offered land of poor quality. During the 2004 PIA study four fully restructured CDFs were studied in detail. On these farms only about 50% of families received land and some former renters even lost land as they were unable to pay for their share title documents. These documents should cost less than two dollars but the CDF administration required much larger sums, reaching 200 somoni ($70) in some cases. To summarise, although reform in the Rasht Valley has been highly variable, a common pattern is one in which a former *sovkhoz* or *kolkhoz* is now split into a few large individual dekhan farms and one CDF, in which a proportion of households are shareholders. Remaining households either rent land from the state fund or have only access to household plots.

In the three mountainous raions of Khatlon Oblast studied, the topography is characterised by low rounded mountains with little permanent snow. Little of the Soviet-period irrigation infrastructure is still functional and thus agriculture is restricted for the most part to rainfed grain production and livestock raising. In this area of the country there is widespread access to land. At the end of 2004, 82% of households reported having access to land other than household plots. However the vast majority of these were renting and only 5% of households had their own individual dekhan farms. By mid-2005 data from the CPS showed that land reform had accelerated and a much large percentage of families had either their own IDF (16%) or titles in a larger CDF (13%), whilst 45% were still renting from an entity they defined as 'the *sovkhoz*'. In 2004–05, with the legal closure of state farms, former *sovkhoz* or *kolkhoz* renters became members of "Renter's Associations" (RA). RA land no longer belongs to the state, nor is it yet registered as an IDF or CDF. Rental contracts are generally short-term (between 1–2 years) and survey respondents underlined the fickle nature of landlord-tenant relations. Some renters are members of state livestock breeding farms which were exempted from the restructuring process and which thus remained as state enterprises rather than being converted to collective dekhan farms.

The study by Action Against Hunger (Porteous, 2003) suggests that in lowland parts of Khatlon Oblast very little individualization of farming has occurred. Only 10% of households have dekhan farms or rented land. Others are either working on state enterprises or unreformed CDFs that operate in exactly the same way as former *sovkhoz*, with workers organized into brigades. Collectives have accumulated large debts from investment companies, which pre-financed cotton production since the mid-1990s and which form part of a chain of intermediaries between the farms and international cotton buyers. At the end of each cropping season, if the returns from cotton were greater than the costs of the inputs provided by the company, then the farm registered a profit, if not it accumulated a debt, which was rolled over year after year. These debts are distributed to both IDFs and CDFs as they are formed. Indebted dekhan farms are forced to buy inputs from investment companies at above-market prices and to sell cotton to them at fixed prices far lower than those of the market. Thus investment companies reap the bulk of the benefit from cotton production in Tajikistan. It is not surprising that very few farmers have taken the steps to register their own dekhan farm. Such farms have no choice over what to grow and have to fulfil their part of fixed government cotton quotas. This fact combined with the inherited debt and the associated forced contracts on poor terms with investment companies mean that many farmers who had acquired private farms now prefer to give them up. Meanwhile those continuing to work as laborers on collectives are often paid only in cotton sticks, burnt as fuel for the winter.

The study in Isfara Raion of Sughd Oblast indicates that here, as in lowland Khatlon, there has been a form of pseudo-privatisation but no individualisation of agriculture. The population density in relation to irrigated land is very high but the climate is good, offering two crops per year. High value crops such as rice, fruit and vegetables are grown for nearby markets in Khojand and the Fergana. In the study area most collective and state farms had been renamed as CDFs. In these collective farms households may sometimes be described as 'share holders', but in reality their true status is that of a farm labourer rather than a private farmer. Land distribution has in fact been little more than a bureaucratic process of assigning workers a theoretical fraction of land (usually between 0.02 and 0.05 ha). Applicants seeking independent tenure may be told by officials that there is no land available or simply receive no reply from the Land Committee. Thus only about 2% of households in Isfara Raion are private farmers in the true sense of the word, but even these may be subject to state production quotas. Ironically those *renting* CDF land are the only ones able to make production decisions for themselves. On collective dekhan farms access to agricultural inputs, irrigation water and machinery is tightly controlled by the CDF head; workers may also rely on him for loans and even for mediation on issues such as electricity access and building permits. This makes secession from the collective all the more difficult. Workers salaries are between about $4.4 and $7.4 per month, but even this may often be paid partly in kind.

4.2. REFORM OF PASTURE LANDS

As mentioned above, the land law makes provisions for long-term heritable use of pasture in much the same way as for arable land. However pasture in Tajikistan is used on a communal basis (Plate 4). In winter and autumn village livestock are grazed together around the village and herding duties are assigned by rota. In summer a few members of the village act as professional shepherds and take all the village animals to the summer pastures (specific areas allocated to each village in the Soviet period), and in return they are paid by other villagers for their services. Thus it is rare to find herds in which all the animals belong to one person. This is a logical consequence of the low stock ownership in Tajikistan where (outside the ethnically Kyrgyz region of Murghab in the Eastern Pamir) households owning more than 50 small stock or 10 head of cattle are rare. For this reason in most parts of Tajikistan the land law is simply not applied to pasture and use of this resource is managed informally at the village level or through farmers' associations such as the case in the Eastern Pamir (Domeisen, 2002). This system promotes the livestock mobility essential to avoid over-use of certain pasture areas and ensures access to remote pastures for all. A recent MSDSP study by one of the authors of this chapter (Robinson 2007) suggests that where summer pastures are fairly local, entailing vertical movements within a single valley or mountain system, migrations have continued. Where movement to allocated pastures entails stock movements across raion or oblast

Plate 4: Summer grazing in Khatlon Oblast. Photograph © Aga Khan Foundation/Jean-Luc Ray

boundaries or horizontal movements of hundreds of kilometres, migration has declined resulting in abandonment of some remote seasonal pastures (see also Domeisen, 2002).

Interviews with the Land Committee and local authorities in GBAO and Rasht in 2004 indicated that they were coming under increasing pressure from central government to distribute pasture equally to households in much the same way as arable land had been distributed, and to ensure that taxes are collected on this land. In Rasht the PIA highlighted how, where land was formally distributed to members, a pasture component was included in the official share given to each former worker.

Due to the herding practices described above, splitting pasture into shares is not practical. Secondly, sharing the pasture tax equally between households with no livestock and those with large herds is likely to be unpopular. The Land Committee in Gorno-Badakhshan and in parts of the Rasht valley indicated that in practice it is probable that pasture shares would not be physically delimited and would exist as such only for tax purposes. Some local governments are looking into the possibility of splitting the overall lump tax payable for each village so that each family pays proportionally for the number of livestock they own, which would be fairer and ensure access to all who need it.

However in some areas permanent land use rights for specific parcels have been granted to individuals on previously communal land. Such is the case in upper Roshtkala Raion in GBAO, where much of the remote summer pasture is now part of individual dekhan farms. The 2007 MSDSP study suggests that this has not so far created conflicts as most of the new landlords do not have large herds and supplement their income by grazing village animals from down the valley, for which they are paid a fee per head. This allows small stock owners continued access to pasture but there is a risk that, as herd sizes rise, the shepherd or his employer will start to exclude communal herds. Results from the PIA suggested that in Khovaling Raion (upper Khatlon Oblast) some pasture lands are controlled by IDF or CDF bosses who then sub-rent these to users. As a result, other livestock owners are often forced into a landlord-tenant relationship with wealthier parties.

These two examples suggest that implementation current land reform laws to pasture could threaten communal management of pasture resources, limit seasonal movements of animals and reduce access for poorer households, potentially leading to a degradation of remaining communal pastures. The slowness with which the status of pastures is being addressed by local authorities may arise from the low pressure on this type of land. However the PIA indicated that livestock numbers are increasing and more families go to summer pastures each year; Robinson (2007) found that some of the longer migrations which had previously been abandoned have recommenced as demand for summer grazing increases. Eventually though, in many areas of the country, the major limiting factor to further growth in livestock production is likely to be winter fodder rather than access to summer pastures.

5. The relationship between land tenure patterns, land quality and income

To summarize, according to land statistics of the government of Tajikistan about 65% of arable land is now in private farms (State Statistics Committee, 2005). However, as we have seen from the examples given above, this is misleading. In reality very few households have their own individual dekhan farm and those who do often privatized large areas through application before 2004, in effect holding more than their theoretical allocation under conditions of equal distribution. Most other households with land access are renting from state enterprises or are shareholders of large collective private farms (CDFs), in some cases without share documents. In the Rasht Valley, even in those situations where land and accompanying documents had actually been shared out to households, a large proportion of families (up to 50%) did not receive land.

Regional differences in the degree and fairness of land reform can be directly related to the quality of the land and thus the vested interests amongst the former *sovkhoz* or *kolkhoz* bosses, and government representatives in keeping control over it. In GBAO small land areas make commercial farming difficult and in Khatlon the lack of working irrigation means that few valuable cash crops can be grown. This helps explain why in those two regions individual access to land is relatively good. Sughd is an irrigable agricultural area close to markets, whilst lowland Khatlon is the major cotton producing area and farming is controlled by the cotton investment companies. Rasht is highly mountainous but availability of irrigable land is good and there is potential to produce apples and potatoes for sale in the capital, thus the situation there is somewhat between these two extremes.

At the household level there are strong relationships between land access and household income. The three MSDSP household surveys measured income including both cash revenue and the imputed value of agricultural production. Table 3 shows patterns of access to land by income group for the Rasht Valley and mountainous areas of Khatlon.

The figures show that households in the poorer quartiles are less likely to have access to dekhan farm land or rented parcels. Even access to presidential land, which was supposed to be targeted at poorer households, follows a similar pattern. Repeatedly, testimony recorded during the CPS and PIA studies confirmed that larger, more productive plots were distributed to well-financed and well-connected individuals, early on in the redistribution process. Figures in Table 3 show that dekhan farms are more likely

TABLE 3. Relationships between land access and income group (source: MSDSP household surveys)

Resource accessed	Group	Percent of households in Rasht 2004	Percent of households in Khatlon 2004**
Dekhan farm arable land*	Richest quartile	25%	11%
	Poorest quartile	4%	0%
Rented farm arable land	Richest quartile	20%	80%
	Poorest quartile	11%	58%
Presidential land	Richest quartile	25%	44%
	Poorest quartile	8%	36%
Irrigation	Dekhan farmers	79%	18%
	Renters	50%	8%

*At the time of the surveys in Rasht and Khatlon almost all households with dekhan farm land had their own IDF. Registration of physical shares in CDFs occurred mostly at the end of 2004 and in 2005 in these regions.

**By 2005 the number of households with IDFs or shares in a CDF was about 18% in the poorest quartile and almost 40% in the richest (CPS data).

than rented shares to be irrigated. They are also much larger in area than rented land (Table 2), again with GBAO being an exception. Kitchen gardens and presidential land tend to be small (0.1 to 0.2 ha), and thus commercial farming is not possible for those owning these tenure types alone. Dekhan farms average over 1 hectare in Rasht and over 4 hectares in Khatlon (Table 2). In conclusion, those who have obtained individual land shares through the reform process have better tenure, larger land areas and are more likely to have access to irrigation than other types of land user. These features are all important for soil management: higher tenure security is likely to encourage investment, it is easier to practice rotation on larger parcels and as we will see later in this paper irrigated land is also more likely to be the object of long term investment than rainfed land. The figures suggest a pattern noted in many areas of the world (Mink, 1993) whereby the poorest farmers, with the least ability to invest, are more likely to be farming the most fragile and marginal lands.

We recognize that there is a causality dilemma related to access to land by the poor. There is evidence that the richest were more able to privatize land in the first place, however whilst the poor have the worst access to land it is also true that those without access to land are likely to be defined as poor measured by the MSDSP surveys in which imputed value of agricultural produce makes up a large proportion of income. The income variable used to define the quartiles includes the imputed value of agricultural produce, which again depends on land area. In a separate paper, the authors investigate this causality more closely by looking at change over time (Guenther *et al.* 2006). Using data from 2001 and 2004, a regression model predicts change in per capita income during this period. We found that change in access to irrigated land was the strongest predictor of change in both total and cash incomes (calculated from the sum of farm produce sold and off-farm income). The positive relationship between land area and income group and the influence of land access on changes in incomes over time suggest that inequalities in land access are one of the major factors behind inequalities in income.

One of the strategies undertaken by the poor to achieve food security under conditions of landlessness is converting sloping rainfed pasture land to agricultural land (Plates 5 and 6). According to MSDSP surveys 5% of families in Rasht admitted to cultivating this 'other land' and the majority were doing so because they had no access to any other land type. In Khatlon about 7% of households were cultivating this type of land, but most of these were also renting land elsewhere. In some areas rainfed land on slopes, not farmed at all in Soviet times, was officially distributed by the local government for use by households for a small fee. An example from Faizabad Raion (Rasht Valley, Plate 7) showed that all households had been using this type of land, although some discontinued the practice after a few years as a lack of fertiliser or crop rotation, combined with intense erosion, lead to low productivity. The evidence suggests that rainfed agriculture generally is being abandoned on marginal land as the returns

Plate 5: Rainfed arable land and pasture above Kevron, Darwaz raion, Gorno-Badakhshan Autonomous Oblast. Erosion following cultivation on steep rainfed slopes can be seen in the background, in contrast with the small irrigated parcel in the foreground. Photograph © Aga Khan Foundation/Jean-Luc Ray

Plate 6: Aerial photograph of rainfed wheat fields near Obigarm/Rogun raions of the Rasht Valley. Photograph © Aga Khan Foundation/Robin Oldacre

Plate 7: Mixture of rainfed and irrigated agriculture in Faizabad raion, lower Rasht Valley. Photograph © Aga Khan Foundation/Jean-Luc Ray

are so poor and other income sources, such as remittances, have increased in importance. In the Rasht Valley the proportion of the sample growing rainfed wheat dropped from 27% to 12% of households between 2001 and 2004, and this decrease occurred mostly amongst those renting land or holding land shares under an insecure tenure arrangement as part of a collective.

6. Agricultural productivity and tenure regime

Agricultural productivity is one indirect indicator of sustainable land management and soil fertility provided that short term productivity increases are not made at the expense of long term fertility. It is of course also affected by many other variables such as seed quality and access to labour and machinery but in the absence of physical measurements of soil quality it is certainly one indicator to monitor. During the 1990s Tajikistan experienced the steepest collapse of agricultural production of any Central Asian country except Kazakhstan. The production index[5] in the year 2000 was at 52% of its 1989–1991 average. Since then Tajikistan has seen quick growth in both GDP and agricultural production, and by 2005 the production index had grown by about 50% on 2000 levels, for both livestock and agriculture (FAOSTAT, 2006). These improvements in productivity are by no means uniform, either geographically or across tenure regimes. Although productivity is affected by many variables, in combination with other data, differences in yield performance may indicate underlying differences in land management and thus it is interesting to compare yield changes on land that is farmed on an individual basis and land that is farmed on a collective basis.

As we have seen, reform in GBAO occurred earlier than in any other region of the country. According to MSDSP and government data, in GBAO from about 1996 onwards wheat and potato yields on the new private farms were more than double those of the remaining collectives and even reached levels well above those on state farms in the late 1980s when the Soviet Union was still functioning (Herbers, 2001). The changes in yield were the result of both tenure regime and an agricultural inputs programme promoted by MSDSP, thus making it difficult to separate the effects of the land distribution itself.

At the national level, Duncan (2000) found that, of all the restructured farm types, only dekhan farms reported an increase in yields (of between 30% and 150%) compared to *sovkhoz* and *kolkhoz*. At that time dekhan farms were mostly the result of individual application and collective dekhan farms were few. As we will see the appearance of collective dekhan farms rather confuses the statistics making it more difficult to compare figures today.

Average productivity figures from 2003 to 2005 comparing 'collective enterprises', 'dekhan farms' and 'household plots' as defined by the State Statistical Committee suggest that yields of staple crops are highest on household plots. They surpass 1991 yields (the last year in which Tajikistan was part of the Soviet Union) on state collectives by 30% and 90% for potatoes and cereals respectively. On collective enterprises equivalent figures are 0%

[5] Sum of price-weighted aggregate volume of agricultural production

(no difference) and 30% respectively. Data for dekhan farms are more difficult to interpret. Yields are as high as those on household plots in the case of potato and as low as those on collectives in the case of cereals. This perhaps reflects ambiguities in the definition of dekhan farms in the statistics[6].

When recent national statistics are used, household plots (which include kitchen gardens and presidential land) are the only unambiguously private and individualized land category in Tajikistan. The World Bank (2006), estimates that half of the 65% growth in monetary value of agricultural sector output between 1998 and 2004 was due to production on kitchen gardens alone. An analysis of crop production shows that output value per hectare sown in 2004 was between 3600 and 3900 somoni for dekhan farms and collectives and 30% higher on household plots (using State Statistical Committee data). However such analyses include areas of cotton and other cash crops which dominate output figures, whilst we know that the 'dekhan farms' growing these crops are in reality indistinguishable from collectives.

Cotton, farmed almost exclusively on collectively organized entities (whatever they may be called) remains at far lower yields than those recorded at the end of the Soviet Union. Average mineral fertiliser application on cotton is the highest in the country (180 kg/ha according to the State Statistical Committee, 2005) as it is provided by investment companies, however total production is still only at 68% of 1991 levels despite no decrease in areas sown to cotton (State Statistical Committee, 2005) and yields are low also in international terms (1.8 mt/ha). Irrigation efficiency is the lowest in Central Asia[7]. Overall 233 thousand hectares, most of which are in cotton growing areas, are subject to high groundwater levels, salinity or both (State Statistical Committee, 2005). This is a severe form of land degradation.

To summarise, there does seem to be evidence that both raw yields of key crops and total productivity in terms of value is higher under the most private and individualised tenure regime and lowest on the most collective and least secure tenure regime. This does not automatically mean of course that there is more land degradation on collectively run entities, but it is sure that crop management is somehow better on household plots and in the sections below we look at some of the reasons for this.

---

[6] Data for 'dekhan farms' are ambiguous as we have seen some are in fact collectively run (CDFs) whilst certain collective enterprises also include individually farmed land as we have seen in Rasht. In addition, outside GBAO, where almost all agriculture is irrigated, it is difficult to interpret data for cereals, as yields on rainfed and irrigated land are mixed in the statistics. Yields will therefore depend as much on the quality of land falling into each tenure type as much as on tenure regime itself.

[7] 125 kg seed cotton per thousand cubic metres of water. In Uzbekistan equivalent figure is 273, in California it is 487 (Goletti and Chabot, 2000).

7. Use of fertiliser and other land management practises in Tajikistan

Tajik farmers are aware of the positive attributes of organic fertiliser application, which promotes good soil structure and increases content of soil organic carbon and essential elements such as nitrogen, potassium and phosphorus. But in much of the study area there are a number of problems with its availability.

Firstly, livestock numbers are low. The average ratio of livestock units[8] per household to hectares of land accessed per household is about 7 in GBAO, 4.5 in Rasht and 2 in Khatlon. Ideally about 20 tonnes of manure is required per hectare and one livestock unit usually produces between 3 and 4 tonnes of manure per year[9]. Table 4 shows rates of organic fertiliser application and relationships with livestock ownership and wealth quartile. In the three regions surveyed frequency of fertiliser use amongst the poor, who own the least livestock, is between 50% and 65% that of the wealthiest quartile.

Secondly, much of the manure produced is not available as people tend to use only the manure from their animal shed accumulated over the winter, while much of what is deposited in the summer or winter pastures stays there.

Lastly, a major factor limiting manure availability is the need to burn it as fuel (Plate 8). The reason for this is that most rural areas have little or no electricity in the winter. Tajikistan does not have enough generating infrastructure and cannot afford to import electricity. Droux and Hoeck (2004) suggest that families will prioritize manure use as fertiliser over its use as fuel, but where

TABLE 4. The use of manure and livestock ownership for various income categories (source: MSDSP surveys)

	Rasht		GBAO		Khatlon	
Quartile	Livestock units per household (avg)	Use of organic fertiliser (% hh)	Livestock units per household (avg)	Use of organic fertiliser (% hh)*	Livestock units per household (avg)	Use of organic fertiliser (% hh)
1(poorest)	0.76	19%	1.54	41%	0.65	34%
2	2.22	19%	3.30	45%	2.0	34%
3	3.6	32%	3.98	58%	3.1	40%
4 (richest)	4.4	37%	4.63	67%	4.9	53%
Total	2.67	27%	3.36	53%	2.67	41%

*Again these figures include Murghab and Khorog, if these areas are not included then rates of use are higher but similar differences between income groups are recorded.

[8] A livestock unit is the equivalent of 1 head of cattle or five head of small stock

[9] Bakhromov, A. and Amirbekov, M. MSDSP Agricultural Department, personal communication.

Plate 8: Manure 'patties' drying in the sun in the Rasht Valley for use as fuel in the winter. Photograph taken by Sarah Robinson

manure is insufficient then alternative sources of fuel must be found. These fuel sources include bushes and trees in Rasht and semi-shrubs such as *Ceratoides papposa* and *Artemesia* spp. (known collectively as *teresken*) in higher areas such as upper Badakhshan and Murghab in GBAO (Plate 9a).

Households in some areas rely on the burning of *teresken* to survive. According to calculations in Domeisen (2002) in Murghab Raion of GBAO, where temperatures may descend to −40°C in winter, a minimum of 112 ha of this bush is cleared every day, an equivalent of about 400 square kilometres every year. In Bartang (upper GBAO) Droux and Hoeck (2004) report catastrophically high rates of *teresken* removal leading to denuded soils and gullying on slopes. The author estimates that *teresken* biomass accumulation rates are about 70 kg/ha per year, an amount burned by a single household in just two days. Collecting *teresken* leads to removal of soil cover and erosion on slopes, and also leads to a decrease in fodder resources for livestock (Plate 9b).

Some NGOs advocate the use of other types of organic fertilisers such as green manure (ploughing of nitrogen rich crop residue into the soil) and household compost. However in many areas all crop residues are fed to livestock in the winter, thus the use of green manure or compost is likely to be an option for households with no livestock or in wetter areas such as Rasht where natural hay can be cut on mountainsides with no irrigation.

Given the above problems and the very high population pressure in Tajikistan, many farmers also use mineral fertiliser either instead of, or as a supplement to, organic fertiliser. Mineral fertiliser, whilst temporarily maintaining levels of nitrogen, potassium and phosphorus (although most farmers

Plate 9a: Spring grazing on semi-shrub vegetation in Murghab raion in the Eastern Pamir, Gorno-Badakhshan Autonomous Oblast. Photograph taken by Sarah Robinson

Plate 9b: Areas close to Murghab town in the Eastern Pamir, Gorno-Badakhshan Autonomous Oblast. Semi-shrubs vegetation has been removed for burning by inhabitants of the nearby town. Photograph taken by Sarah Robinson

apply nitrogen only) does not contribute to maintenance of soil structure and moisture holding capacity. Whilst in itself not harmful if used in the correct quantities, its use may lead farmers into abandoning practices which assure longer term soil quality, such as organic fertiliser application and crop rotation. However, we have seen that there are limits to the amount of organic fertiliser available and mineral fertiliser at least prevents soil mining in which the soil is completely depleted of all the minerals required for plant growth.

In Soviet times the use of mineral fertilisers was high as they were provided by the state. Use plummeted during the economic collapse of the 1990s; the lowest year was 1999 when fertiliser consumption was less than one sixth of 1992 levels. By 2002 consumption had reached 40% of 1992 levels (FAOSTAT, 2006). In 2002 just under half of all nitrogenous fertilisers used were produced in country (FAOSTAT, 2006), and the rest were imported from Uzbekistan or Russia. Today this input may be found on the market in most areas, although there are exceptions such as the remote raion of Shurabod in upper Khatlon Oblast, however it is expensive and many farmers complain that they cannot pay for fertiliser in cash during the sowing season. Unsolicited responses from focus groups in all three raions of Khatlon Oblast suggest that limited ability to purchase fertiliser has been a major contributing factor to poor yields.[10] As we will see in the next section, access to agricultural credit greatly increases mineral fertiliser use and goes a long way to explaining regional differences in the application of mineral fertiliser.

7.1. APPLICATION OF MANURE AND FERTILISER BY CROP TYPE AND TENURE REGIME

Maintenance of soil fertility and prevention of land degradation depends on farmers' ability to conserve arable and pasture land though sustainable practices. Table 5 presents data for mineral and organic fertiliser use for different crops and land tenure types across the three regions. The data suggest that there is a tendency to use mineral fertiliser more on private or rented land shares and manure on kitchen gardens. This is perhaps due to the larger size of private and land shares and their physical distance from animal quarters near kitchen gardens where manure is accumulated. There are large numbers of farmers who are not adding organic matter to their land shares. The addition of mineral fertiliser without organic matter leads to a decrease in soil organic carbon. Most worrying is the fact that a significant percentage of farmers use no soil improvement methods at all.

[10] During the CPS respondents in all three raions also mentioned the lack of soil conservation methods and lack of training in such methods as detrimental to soil quality.

TABLE 5. Data on fertiliser and manure use by private farmers in Rasht, Khatlon and GBAO regions. (Source: MSDSP household surveys)

	Rasht		GBAO		Khatlon	
Agricultural practice	Sample size	Fertiliser use	Sample size	Fertiliser use	Sample size	Fertiliser use
Planted potato off household plots	145	**48% total** mineral 41% organic 20%	544 (both land use types)	**85% total** mineral 56% organic 60%	76	**55% total** mineral 30% organic 30%
Planted potato on household plots	606	**55% total** mineral 31% organic 37%			390	**72% total** mineral 21% organic 67%
Planted irrigated wheat	108	**58% total** mineral 56% organic 4%	398	**86% total** mineral 73% organic 36%	33	**30% total** mineral 21% organic 12%
Planted rainfed wheat	114	**25% total** mineral 21% organic 8%	54	**10% total** mineral 10% organic 0%	762	**12% total** mineral 10% organic 3%
Bought mineral fertiliser (out of total sample)	867	33%	691	59%*	1000	18%
Used organic fertiliser (out of total sample)	867	27%	691	53%*	1000	41%

*If Khorog city and Murghab (where agriculture is not practiced) are not included, the access figures are 69% for mineral fertiliser and 60% for organic fertiliser.

The data also suggest that fertiliser of any type is more likely to be used on irrigated land than rainfed land. This is partly because once the soil is dry fertiliser is less likely to be available to the plant. However Tajik agronomists have shown that in rainfed areas such as those found in Rasht and Khatlon the application of nitrogenous fertiliser in the spring increases yields[11]. Thus economic factors, mainly the cost of fertiliser in relation to benefits, probably explain low fertiliser use on rainfed land.

As we have seen, on former collective land (outside household plots) there are a number of degrees of security of tenure. These were also compared in terms of fertiliser utilisation. In Rasht the frequency of application of fertiliser (all types) or mineral fertiliser alone on irrigated potato was almost the same for the two groups of dekhan and renting farmers. This suggests that there is little difference between these two tenure types in terms of the willingness to invest in mineral fertiliser, perhaps because the benefits of adding fertiliser last for one year, during which tenure is secure. In the Rasht Valley, MSDSP reports a large number of renting farmers amongst clients of their agricultural credit programmes; in 2002 40% of participating farmers were renters.

As well as differences in the frequency of mineral fertiliser use between tenure or crop types the data reveal large differences between the three regions. In GBAO use is 69% (of households in agricultural areas), in Rasht 33% and Khatlon 18%. Whilst it is possible that good tenure security in GBAO may be important, another reason is likely to be the availability of agricultural credit. MSDSP survey data showed that only 8% of households had access to agricultural credit in Khatlon, mostly provided by friends and family, but NGO programmes have made credit available to 18% of households in Rasht and 47% of farmers in GBAO. It should be stressed that high reliance on mineral fertiliser is not the single solution to long term conservation of soil fertility. In GBAO its use is held artificially high by the fact that credit is subsidised and in-kind repayments were accepted for many years. Much agriculture in GBAO is subsistence in nature, and as the subsidy has been cut back and cash repayments introduced, fertiliser use has started to drop, exposing the lack of more sustainable ways of soil fertility conservation such as organic fertiliser application, rotation and physical protection such as terracing.

7.2. OTHER SUSTAINABLE LAND MANAGEMENT PRACTISES

Fertiliser application (in particular organic) is only one factor important for conservation of soil fertility. Crop rotation is another. Quantitative data are not available but interviews with farmers between 2003 and 2006 suggest that

[11] Amirbekov, M., MSDSP head agronomist, personal communication.

rotation on irrigated areas tends to be between potato and grain crops only. Although some farmers plant leguminous fodder crops such as lucerne and sainfoin, the primary aim is fodder production rather than crop rotation. These crops are highly productive, providing 4–5 cuts per year, but lack flexibility for rotation as they are perennials and are usually cultivated for 5–10 years on the same parcel. Seed is extremely expensive (10–15 somoni/kg compared to 1 somoni/kg for wheat seed) and this is another reason why farmers prefer varieties which persist for many years. Farmers commonly state that in order to switch to alternative annual legumes then these would have to be cheap to plant and as productive as lucerne, but such criteria will be difficult to meet. Intercropping with leguminous fodder crops or their use as autumn cover crops could also be tried in order to return nitrogen to the soil without the need to sacrifice areas planted to food crops. A number of bean and pulse varieties are cultivated but the market potential is limited as they are not staples in Tajik cuisine. In upper areas of Gorno-Badakhshan wheat is often intercropped with beans, but these are for home consumption only.

Terracing for erosion control is another possible way to improve soil fertility. In some parts of Tajikistan, such as Gornaya-Matcha Raion in Sughd Oblast, terracing is now used extensively and in GBAO its use is increasing with NGO support. However in both areas terraces are built only on irrigated land. Erosion and loss of soil fertility are probably greatest on rainfed land as this is often located on steep slopes and, as we have seen, little fertiliser is used on such land. However rainfed land is almost never terraced; investments in labour and capital are too great given the low and erratic returns. Terraces for fruit trees were built in the Soviet period on rainfed land around Dushanbe and in mountainous parts of Khatlon Oblast, where rainfall is relatively high, but the trees have to be irrigated from water tankers for their first few years, something which is not economically feasible today. Overall communities are very keen to make investments in irrigation and terraces on irrigated land, and this does have the advantage of increasing the productive area and making cultivation of rainfed slopes less attractive.

TABLE 6. Yields of crops in each income quartile as a percentage of yields in quartile four (source: MSDSP surveys).

	Rasht			Khatlon		GBAO	
Quartile	rainfed wheat	irrigated wheat	potato	rainfed wheat	potato	irrigated wheat	Potato
1 (poorest)	84%	33%	43%	88%	67%	76%	79%
2	71%	64%	62%	96%	69%	90%	86%
3	76%	94%	89%	96%	77%	97%	97%
4 (richest)	100%	100%	100%	100%	100%	100%	100%

7.3. SUSTAINABLE LAND MANAGEMENT AND INCOME

Soil conservation and improvement practises will always be hardest for the poorest income group to implement. Data from MSDSP surveys (Table 6) showed that in mountainous parts of Khatlon and GBAO this group has average yield rates for staple crops of between 70% and 80% of those of the wealthiest group. In Rasht, where land distribution was least equitable and income inequality (measured by the Gini coefficient) is highest, average yields of irrigated potato and wheat amongst the poorest quartile were between 30% and 40% of those obtained by households in the wealthiest quartile. Yields on rainfed land tend to be less variable than those on irrigated land, with the poorest quartile achieving over 80% of the wheat yields of wealthiest quartile. This suggests that investment in rainfed land is low across all income groups. Yields are related to farming knowledge and experience, pest control, seed type and other variables not related to soil quality. But data presented here do suggest that there are large differences in organic fertiliser application between income groups and MSDSP surveys also show a similar pattern for mineral fertiliser application. Outside GBAO wealthier households were more likely to obtain the best and largest parcels of land. It is thus probable that these factors do play at least some role in yield difference and soil management between income groups.

8. Conclusions

The literature suggests that farmers with access to their own plots of land and secure land tenure are most likely to invest in land improvements and long term soil fertility. From this study we can make the following observations on the state of land reform in Tajikistan.

- In mountainous areas reform has led to both privatization and an individualization of farming but although farming is now undertaken at the household level, these households are often renters or shareholders in 'collectives' whose managers still have much control over the land.
- Poorer households generally have the least secure tenure arrangements, worst quality land and smallest land areas. Some households did not receive any land at all during the reform process. Such groups are the most likely to cultivate rainfed land on steep slopes, which looses fertility and quickly goes out of production.
- In productive areas of the lowlands there has been little land distribution to households and much farming is still conducted by labourers working on collective farms.
- Access to pasture is generally good but some remote pastures have been abandoned due to risks and costs association with travelling to them,

increasing pressure on pastures near villages. Most pastures are still managed as common property but this system is presently threatened by legislation allowing individuals to apply for permanent use of delimited pasture shares. This could result in ecosystem fragmentation, a reduction in livestock mobility and loss of pasture access for village herds. Local authorities are searching for ways to reconcile the traditional norms of pasture use with the land law, and thus the direction that pasture management will take is still unclear.

These reforms have lead to the following patterns in agricultural productivity and investment in soil fertility:

- There is some evidence that agricultural productivity is highest on the most 'private' form of tenure regime which is that of the household plot. Yields are lowest on collective 'state enterprises'. On dekhan farms the data show a mixed picture, perhaps as some dekhan farms are actually collectively run and thus this category is ambiguous. But in the case of crops such as potato, a cash crop usually planted by individual households, yields on dekhan farm land are as high as on household plots.
- Household survey data show that those in the lowest income quartile have the lowest rates of fertiliser application and the lowest crop yields in all three regions studied. Such households will be the least likely to invest in sustainable soil management.
- Where cotton is grown, state plans and high levels of debt mean that crop rotation is almost never used, cotton is planted year after year using high levels of mineral fertiliser. This highly intensive farming combined with inefficient irrigation has led to high levels of soil salinity.
- Outside cotton growing areas organic and mineral fertilisers are most likely to be used on household plots (small, close to the household and tenure-secure) but their use is less frequent on other types of land. Fertiliser of any type is rarely used on rainfed land and such land looses fertility quickly, going out of production in some areas.

Tajikistan suffers from a number of serious structural problems such as a lack of electricity and high population growth, which are the root causes of land degradation. The biggest land degradation threat comes from fuel burning (Plate 10). Collection of shrubs leads to erosion and loss of pasture plants for grazing. In the Eastern Pamir this has reached catastrophic proportions. Manure use is limited by the need to burn it as fuel and crop rotation hindered by small parcel size and high cost or low availability of appropriate legume seed. High population growth and lack of alternatives to agriculture hinder farm consolidation and leave many households in a subsistence trap, unable to accumulate capital to make long term investments in their land.

Plate 10: Girls returning from collecting firewood in the Bartang Valley, Gorno-Badakhshan Autonomous Oblast. Photograph © Aga Khan Foundation/Robert Middleton

8.1. RECOMMENDED LEGISLATIVE OR POLICY CHANGES

A number of policy changes could be made which might help ease the painful process of agricultural transition in Tajikistan and reduce poverty while favouring sustainable land management:

- Increasing land access to individual households would probably encourage productivity and reduce poverty. In lower Khatlon Oblast this would require confronting the vested interests associated with cotton production, interests which have been immune to pressure from international organisations for many years.
- The literature suggests that farmers with secure tenure are more likely to invest in land improvements and long term soil fertility. In Tajikistan reducing the powers of government to confiscate land or transfer it to another party would go some way to improving this security. Abolishment of the CDF structure, implying distribution of certificates to all current share holders, would also be an improvement.
- The ability to sell land would clearly enable those families who want to leave farming to do so with some kind of financial gain. At the same time land sales would allow farms to consolidate and encourage investments in soil conservation.
- Pasture tenure laws are at present ill-adapted to communal forms of grazing. The pasture tax should be set per head of livestock rather

than asking each share-holder to pay tax on an equal area of pasture regardless of the number of animals owned. This would allow for flexible management of livestock movement at the village level depending on season and livestock numbers and enable the continuation of shared herding.

8.2. PROPOSED CHANGES TO THE LAND LAW CURRENTLY UNDER DISCUSSION

In 2007 a number of amendments to the Land Code and Law on dekhan farms were proposed. Key changes include the introduction of land markets, removal of undefined reasons for land withdrawal such as non-rational use and introduction of stated rights of farmers to choose the crops which they grow. These changes have been proposed by the Working Group on Structural and Land Reform which includes actors from both government and international organisations such as the World Bank. Many of the changes to the land code were watered down or removed as the text went to Parliament in 2007, and are still under review by parliamentary committee at the time of writing. There are at least two draft versions of the law on dekhan farms discussions on both continue. So far no changes regarding pasture tenure have been proposed.

Acknowledgements

A Study supported by the Aga Khan Foundation and Mountain Societies Development Support Programme, 2006. Disclaimer: the opinions and statements expressed in this paper do not represent those of the Aga Khan Foundation or its partner organizations.

References

Domeisen, M., 2002, *Marginalised by the Impacts of Transformation. A study of post-Soviet livestock breeding in the high mountains of the Eastern Pamirs*, Diploma Thesis submitted to the faculty of Natural Sciences, University of Berne.

Droux, R., and Hoeck, T., 2004, *Energy for Gorno-Badakhshan: hydropower and the cultivation of firewood*, Joint Diploma Thesis submitted to the faculty of Natural Sciences, University of Berne.

Duncan, J., 2000, Agricultural land reform and farm reorganization in Tajikistan, *Rural Development Institute Reports on Foreign Aid and Development,* **106**: 1–46.

FAOSTAT, (2006); http://faostat.fao.org/faostat/.

Giovarelli, R., 2004, *Land Legislation in the Republic of Tajikistan*, A report for the ARD/CHECCHI Commercial Law Project.

Goletti, F., and Chabot, P., 2000, Food policy research for improving the reform of input and output markets in Central Asia, in: *Food Policy Reforms In Central Asia: Setting the*

Research Priorities, S. Babu, and A. Tashmatov, ed., International Food Policy Research Institute, Washington, D.C.

Guenther, T., Robinson, S., Jumakhonova, R., and Otambekov, A., 2006, *Moving out of poverty in rural Central Asia: long term economic development or high income volatility?* Analytical paper for the Aga Khan Foundation.

Herbers, H., 2001, Transformation in the Tajik Pamirs: Gornyi-Badakhshan—an example of successful restructuring? *Central Asian Survey*, **20**(3): 367–381.

Land Committee of the Government of Tajikistan, 2004, *Cvod Zakonov I Drugie Normativno-pravovye Akty o Zemle* (*The Land Code and Other Legislation on Land*), Sarparast, Dushanbe, Tajikistan.

Lerman, Z., Csaki, C., and Gershon, F., 2002, *Land Policies and Evolving Structures in Transition Countries*, World Bank.

Lerman, Z., 2000, From common heritage to divergence: why the transition countries are drifting apart by measures of agricultural performance, *American Journal of Agricultural Economics*, **82**(5): 1140–1148.

Macours, K., and Swinnen, J. F. M., 2000, Impact of initial conditions and reform policies on agricultural performance in Central and Eastern Europe, the Former Soviet Union, and East Asia, *American Journal of Agricultural Economics*, **82**: 1149–1155.

Mink, S., 1993, *Poverty, Population and the Environment*, World Bank Discussion Paper, 189.

Neef, A., 2001, Land tenure and soil conservation practises-evidence from West Africa and South East Asia, in: *Sustaining the Global Farm, Selected Papers From the 10th International Soil Conservation Meeting*, D. E. Scott, R. H. Mohtar and G. C. Steinhardt, ed., May 24–29, Purdue University and the USDA ARS National Soil Erosion Research Laboratory.

Olimova, S., and Bosc, I., 2003, *Labour Migration from Tajikistan*, International Organisation for Migration/Sharq Scientific Research Centre.

Ostrom, E., 1990, *Governing the Commons: the evolution of institutions for collective action.* Cambridge University Press, UK.

Porteous, O., 2003, *Land Reform in Tajikistan: From the Capital to the Cotton Fields*, Report for Action Against Hunger.

Robinson, S., 2007, *Pasture management and Condition in Gorno-Badakhshan: A Case Study*. Report on research conducted for the Aga Khan Foundation Tajikistan, http://www.untj.org/library.

Rohde, R. F., Moleele N. M., Mphale, M., Allsopp, N., Chanda, R., Hoffman, M. T., Magole, L. and Young, E. 2006, Dynamics of grazing policy and practice: environmental and social impacts in three communal areas of southern Africa. *Environmental Science and Policy*, **9**, 302–316.

Southgate, D., 1988, *The Economics of Land Degradation in the Third World*, World Bank Environment Department Working Paper, **2.**

State Statistical Committee of Tajikistan, 2005, *Sel'skoe Khozyaistvo Respubliki Tadjikistan* (*Agriculture in the Republic of Tajikistan*), Dushanbe, Tajikistan.

UNDP, 2005, *Moving Mountains, the UN Appeal for Tajikistan.*

Wen, J. L., Saleem, H. A. and Qian, Z., 2007, Property rights and grassland degradation: a study of the Xilingol Pasture, Inner Mongolia, China, *Journal of Environmental Management*, **85**, 461–470.

World Bank, 2006, Priorities for Sustainable Growth: a strategy for agriculture sector development in Tajikistan.

World Bank (February 9, 2006); http://siteresources.worldbank.org/INTTAJIKISTAN/Resources/TASS-F_R_V.pdf.

CHAPTER 9

ISRAELI EXPERIENCE IN PREVENTION OF PROCESSES OF DESERTIFICATION

PREVENTING DESERTIFICATION IN ISRAEL

N. ORLOVSKY*

Department of Dryland Biotechnologies, The J. Blaustein Institutes for Desert Research, Ben-Gurion University of the Negev, Sede Boqer Campus 84990, Midreshet, Ben-Gurion, Israel

"We must conquer the desert, lest the desert conquers us" Ben-Gurion.

Abstract: Since over 60% of Israel is occupied by the Negev desert, measures to combat desertification were initiated at an early stage of the country's development and have intensified. In the agricultural sector, substantial savings have been achieved through technological improvements in irrigation methods, increasing the efficient use of water and effluents, promoting water recycling, minimizing pesticide use, advancing organic and greenhouse agricultures, and the development of new crops and innovative machinery. Scarcity of water, limited land resources, and lack of natural resources have led Israel to base its economy on technological advances that have considerably reduced the risk of desertification.

Keywords: combating desertification, Negev desert, runoff agroforestry, organic agriculture, drip irrigation

1. Geographical features

Israel is located on the eastern shores of the Mediterranean Sea in the South West corner of Asia on the driest border of the Mediterranean climatic zone. The total land area of the country is more than 21,000 square kilometers. Long and narrow in shape, the country is about 470 km in length from north

* To whom correspondence should be addressed. Department of Dryland Biotechnologies, The J. Blaustein Institutes for Desert Research, Ben-Gurion University of the Negev, Sede Boqer Campus 84990, Midreshet, Ben-Gurion, Israel

R. Behnke (ed.), The Socio-Economic Causes and Consequences of Desertification in Central Asia, 205–229.

to south and some 135 km across at its widest point between the Dead Sea and the Mediterranean coast. Located at the crossroads of Asia, Africa and Europe, the country makes up for its small size with a wide range of ecological conditions. The Implementation Annexes of the UN Convention for Combating Desertification refer to four regions of the earth that suffer major problems of desertification: Three of the four meet in Israel. The southern part of the country joins the Sahara-Arabian desert belt. The central part is an extension of the Irano-Turanian region, and northern Israel borders the Mediterranean region.

Israel's climate has unique characteristics and ranges from temperate to tropical, with much sunshine. One of the main characteristics of this kind of climate is the high variability in precipitation from year to year and between different areas. In the southern most area, the amount of rainfall is as low as 25 mm per year, while in the north of the country it is up to 950 mm per year (Fig. 1). Generally, Israel has between 8 to 75 rainy days annually, mainly in the period from October through April. There is a clear division into two seasons: a hot and dry summer and cool rainy winter.

Some 60% of the country is occupied by the Negev desert, which is comprised of three out of the four dryland types – hyper-arid, arid and semi-arid. It is a tiny desert by world standards but its empty areas are Israel's open frontier for sustainable development.

The semi-arid northern Negev, with a mean annual precipitation of 200–350 mm, supports livestock breeding, rainfed winter wheat sometimes supplemented by irrigation, and other irrigated croplands. The arid Negev highlands, with up to 200 mm of rainfall, are characterized by vast barren areas, some of them irrigated and partly used as natural rangelands. The hyper-arid southern Negev and Arava Rift Valley, with up to 50 mm of rainfall, are mostly barren, with some irrigated oasis agriculture.

Since the establishment of the State of Israel an intensive program of research and development, coupled with resettlement projects, has transformed large sections of this region into vitally important productive areas, contributing to the progress and well-being of the entire nation. Unique characteristics make the Negev desert suitable for producing out-of-season crops and for tourism, hiking, four-wheel driving, and rangeland grazing. Now only 6% of the country's residents live there; 33–36% of the Negev's area is occupied by military training zones and nature protection departments; 10% of the area is used for rainfed agriculture and grazing, and only 6% of the land is irrigated. Finally, 3%, 2% and 1% of the land are under urban use (including roads), landscape enhancement or mining (Fisher et al., 1999). Combating desertification is based on the rational use of soil and water resources, runoff agroforestry, controlled-environment desert agriculture, organic agriculture, environmental protection and nature conservation.

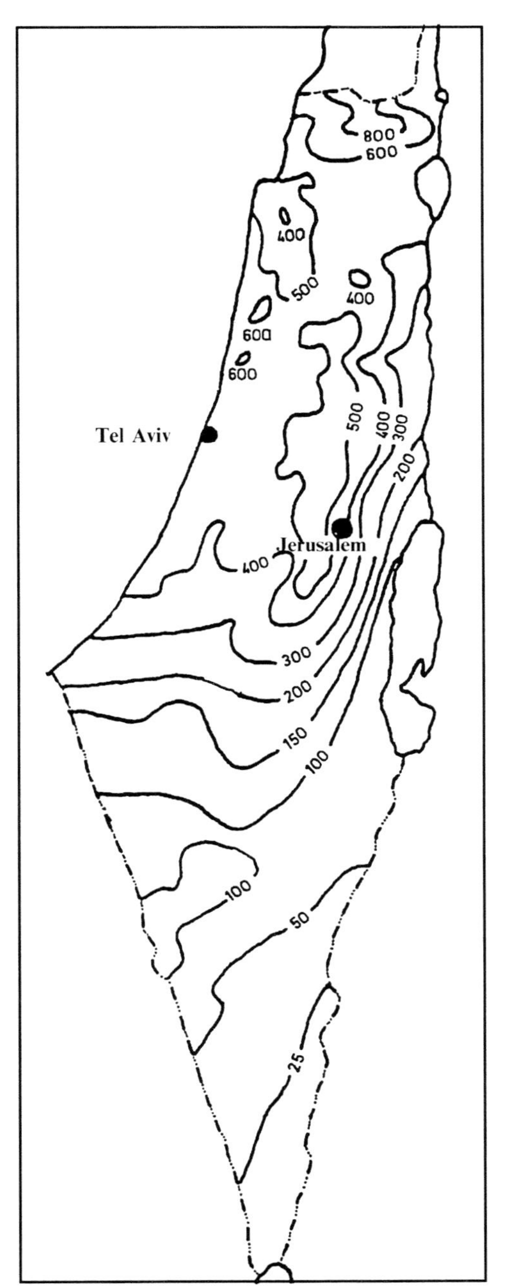

Figure 1. Distribution map of annual precipitation in Israel (Atlas of Israel)

The obvious direction for recovery of arid areas is the supply of additional water and more efficient use of already available supplies. This is essentially the major policy that has guided Israel in bringing its arid areas into productive use. This task was undertaken as a concerted effort, combining technological means with social and economic methods and remarkable results have been achieved over the last 30 years.

2. Water as a major factor in the recovery of degraded arid ecosystems

Israel is learning to live in and with the desert rather than to exploit it to the point of non-sustainability. The first and most important consideration is water. The available water in Israel averages about 2,000 million cubic meters (MCM) annually. About two-thirds of Israel's annual fresh water potential is derived from the three major sources: Lake Kinneret, the coastal aquifer, and the inland mountain aquifer. Lake Kinneret, Israel's only natural surface storage reservoir, has an utilizable water yield of about 512 MCM, the mountain aquifer has total safe yield of about 326 MCM while the coastal aquifer has total safe yield 418 MCM. The rest is made up equally from smaller aquifers especially in the Western Galilee and the Arava/Negev region (337 MCM), and from recycled and brackish water (333 MCM) (Lithwick, 1998). These sources are dependent primarily on annual additions through rainfall, which is problematic on a variety of grounds, the most important being short-term climatic variability and the possibility of longer-term periods of significant declines in the form of epochs of drought. The Lake Kinneret has had annual inflows ranging from a low of 100 MCM in drought years to a high of 1,500 MCM (Kliot, 1994). These phenomena impose on planners the need to make appropriate risk allowances in estimating future requirements.

Because no more readily-used water resources were available for development, efforts to increase water potential have led to exploitation of marginal water resources. These include treating wastewater, storing runoff, the use of brackish, geothermal fossil and sea water. By 1999, total water resources amounted to 2,151 MCM – 52% fresh groundwater and springs, 31% surface water from the Jordan watershed, 12% marginal water, and 5% flood water. The increase of 151 MCM came mainly from the treated wastewater and desalinization (Statistical Abstract of Israel, 2000).

Israel is a world leader in recycling wastewater, which now accounts for 20% of total supply, up from 3% two decades ago. Nearly 70% of the wastewater collected in sewers is treated and reused for agricultural purposes, mainly for the irrigation of non-food crops and animal fodder. The treated water is recharged into the coastal aquifer; it is then pumped and transferred by a special pipe to the northern Negev. Recycled wastewater may theoretically provide 400 MCM of water per year for irrigation purposes (The Environment in Israel, 1999).

Israel already uses some 180 MCM of brackish ground water a year for agricultural and industrial purposes. Potential of brackish ground water in northern Israel amounts to about 230 MCM; however, in southern Israel, its potential is much greater and amounts to a billion cubic meters.

This water is used directly by the Dead Sea industries (18 MCM) and for irrigation in the Arava Valley (12 MCM) (Nativ and Iassar, 1989; The Environment in Israel, 1994).

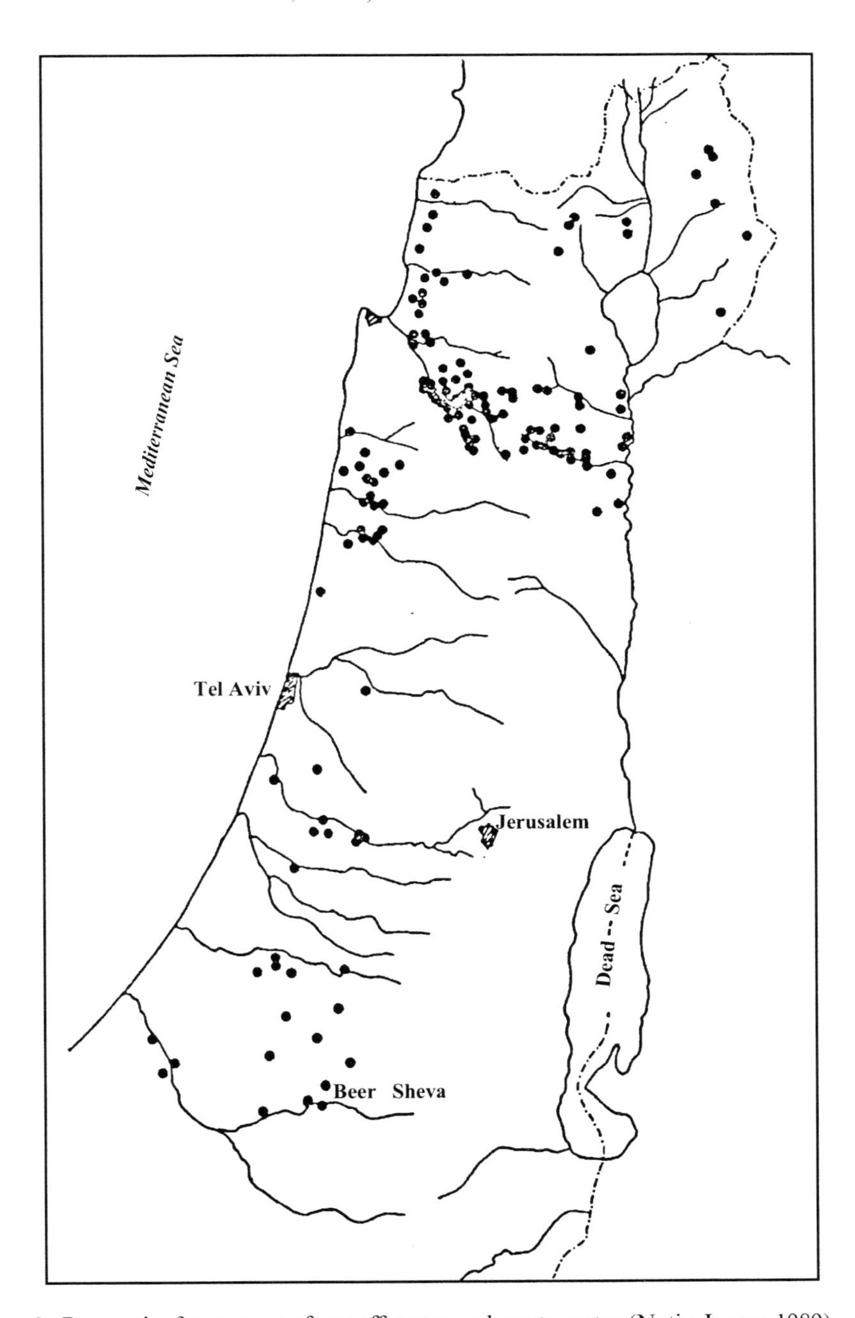

Figure 2. Reservoirs for storage of runoff-water and waste-water (Nativ, Iassar, 1989)

Brackish water can be desalinated. In Israel, nearly 40 desalination units, used for water supply and research purposes, have been built over the past few decades - with a total capacity of 50,000 cubic meters per day. Since 1979, water for Eilat at the southern tip of Israel has been partially supplied by several desalinating units. In January 1994, a new desalination plant was inaugurated in Eilat with a production capacity of 6,300 cubic meters per day, raising the capacity of all desalination facilities in this area to 27,000 cubic meters per day.

Over 260 surface reservoirs were constructed alongside or on ephemeral streams to harvest flash-floor water that would otherwise be lost (Fig. 2). Usually, these reservoirs contain also treated waste water. Their volume ranges from 0.01 to 6.5 MCM, with a mean volume of about 0.6 MCM. Their total capacity amounts to about 150 MCM (Nativ and Iassar, 1989).

The government has prepared a plan for the economic use of water and to overcome any water crisis up to 2010. Its main aim is to stabilize the water economy by replenishing the reservoirs and improving water quality, establishing additional reservoirs for rain and floodwater, cutting freshwater allocations to agriculture to 530 MCM, increasing desalination to 400 MCM of water per year, and increasing investments in wastewater treatment for agriculture, gardening and landscaping at a scope of 230 MCM per year (The Environment in Israel, 2002).

3. Consumption of natural resources

The overall water plan for Israel integrated all the resources into a comprehensive scheme so that (a) water could be conveyed from regions of excess in the north to regions of scarcity in the south and (b) the system would have operational flexibility so that surface water and groundwater could be stored in aquifers and transferred interregionally (Shanan, 1998). This storage transfer and distribution of water is carried out by the National Water Carrier (Fig. 3). The Carrier is a combination of underground pipelines 70 to 106 inches in diameter, open canals, interim reservoirs, tunnels and pumping stations, supplying about 400 MCM annually from Lake Kinneret, located 220 meters below sea level. Water is pumped to an elevation of about 152 m above sea level, and flows by gravitation to the coastal region, whence it is pumped to the Negev. In addition to the Lake Kinneret, two large aquifers, the Mountain Aquifer and the Coastal Aquifer, contribute both about 600 MCM per annum to the Carrier. Construction commenced in 1953 and was completed in 1964. The length of the system's main conduit is 130 km. The National Water Carrier functions not only as the main supplier of water, but also as an outlet for surplus water from the north in winter and early spring and a source of recharge to the underground aquifers in the coastal

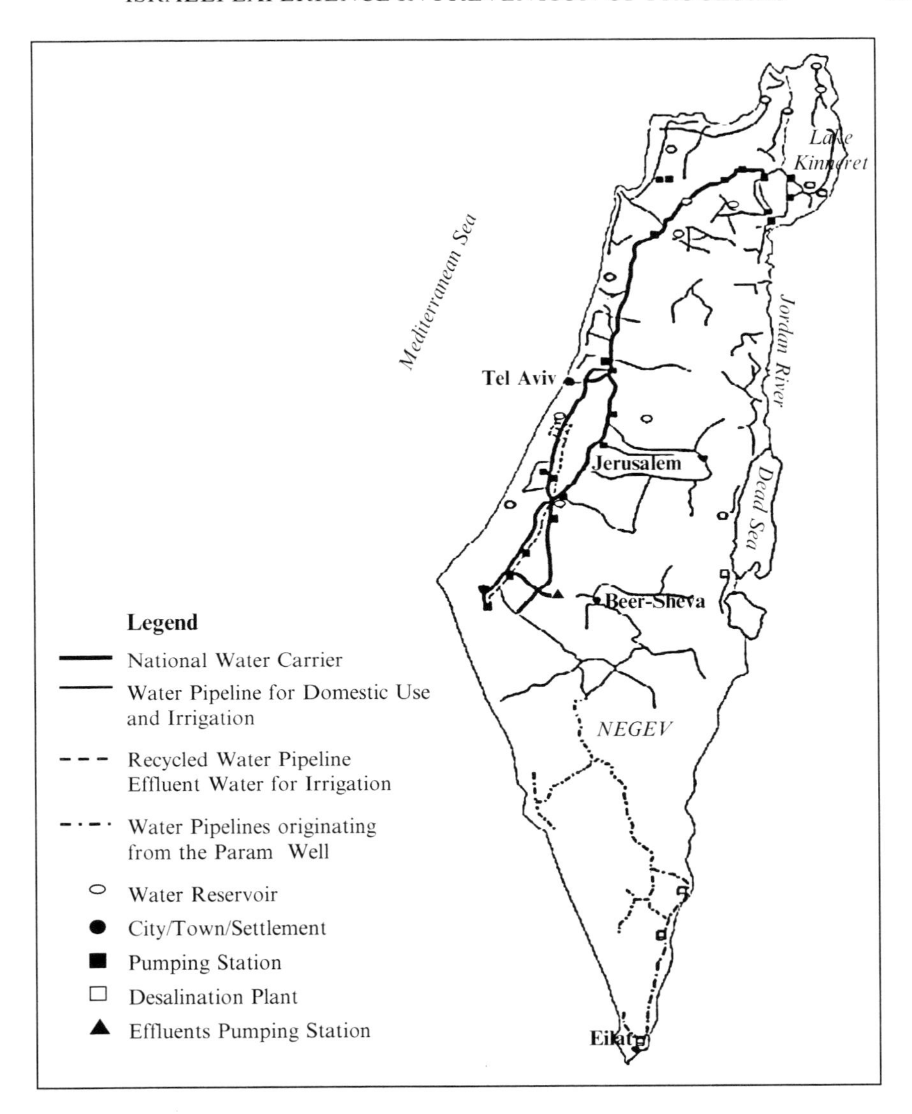

Figure 3. Water supply system of Israel (Sitton, 1998)

region. Most of the regional water system are incorporated into the National Water Carrier to form a well-balanced network in which water can be shifted from one line to another according to conditions and needs (Sitton, 1998).

Present estimates indicate that Israel currently uses as much as 95% or more of its total renewable water resources, including both surface and groundwater. The main consumers of water are the agriculture, domestic and industrial sector. The industrial sector has increased consumption of water with 55 MCM in 1965 up to 122 MCM in 2002 that corresponds from 4 up to 7% of the general consumption of water. The consumption of water has increased with 199 MCM up to 688 MCM in domestic sector for this period. Water use in the agricultural sector fell from over 80% of total to about 50% of total use between 1948 and the late 1980s and made 1,264 MCM.

In the twelve years following the establishment of the State in 1948, cultivated land area increased from 160,000 to 400,000 hectares – about a fifth of Israel's land area (Statistical Abstract of Israel, 2002). Thereafter, cultivation increased far less rapidly, stabilizing 440,000 hectares in the early 1980s, while shifting somewhat from central Israel to peripheral areas. The arable land for one person gradually decreased with 2,500 square meters in 1951 up to 750 in 1995 and 650 square meters in 2000. Water scarcity is the main limiting factor in Israel agriculture and the country depends on irrigation to increase its crop yields. Irrigated land has increased since 1948 from 30,000 to 192,100 hectares in 2002 (Statistical Abstract of Israel, 2002).

For prevention of ecosystems degradation the government of Israel emphasizes improving the efficient use of water and effluents and promotes water recycling, minimization of pesticide use, advancement of organic agriculture, and development of new crops and innovative machinery.

4. Desertification control in agriculture

The lack of water is the most severe constraint on Israeli farmers. This deficiency influences agriculture in two ways: first, by limiting the amount of land which can be cultivated, and second, by inducing farmers to use both land and water as efficiently as possible. Toward this latter goal, highly mechanized, high-input methods and water-saving irrigation systems are employed. Today Israeli agriculture is characterized by a high technological level and is based on micro-irrigation systems (drip, mini-sprinkler and underground irrigation), automatic and controlled mechanization, and high quality seeds and plant.

Efficient use of water is crucially dependent on advanced irrigation technologies especially in arid-land farming. Until about 50 years ago crops in Israel were irrigated by surface (flood and furrow) irrigation. Surface irrigation is possible only when the ground is leveled and the soil type enables slow or moderate percolation of the water. Under arid conditions, surface irrigation leads to severe loss of water by evaporation and to percolation beyond the developed root system, especially in the stage of germination and early development. Moreover, between irrigation sessions the plants are exposed to stress. Another negative aspect of surface irrigation under arid and semi-arid conditions is the process of soil salinization. Vast areas in the arid and semi-arid regions affected by salinization have indeed had to be abandoned.

Pressurized irrigation with sprinklers, introduced in Israel since the mid-1960s, contributed much to modernizing agriculture and increasing water use efficiency. However, from the standpoint of agriculture in arid and semi-arid regions, the most important development has been the introduction of drip irrigation. Drip irrigation has been developed in Israel and introduced into Israeli agriculture less than 30 years ago. At present, over 140,000 hectares of irrigated area in Israel use drip irrigation (The Environment in Israel, 1992). Drip irrigation has many advantages over other irrigation methods (Sitton, 1998):

- Drip irrigation is the most efficient method of irrigation when it comes to conserving water. Since the drippers emit the water directly to the soil adjacent to the root system, which absorbs the water immediately, evaporation is minimal. This characteristic is especially important in arid zones. By sprinklers or surface irrigation methods, evaporation is enhanced by wind, while in drip irrigation the impact of wind is minimal
- Water is discharged uniformly from every dripper fitted onto the lateral pipe. This is true even on moderately sloping terrain. Furthermore, the development of compensated drippers enables uniform irrigation on steeper slopes over greater distances.
- Via the drippers, fertilizers can be supplied to the plant together with the water.
- The quantity of water delivered can be optimized to fit different soil types, avoiding percolation of water beyond the root zone.
- The emergence of weeds is minimized.
- Drippers with a given discharge of water (of the order of several liters per hour) can be installed at any spacing to accommodate the needs of any crop.
- Exploitation of poor quality water (saline water or effluents) is made possible.
- Saline water can be used because direct contact between water and leaves is avoided, thus obviating burns.
- Drip irrigation causes salts to be continuously washed away from the root system, avoiding salt accumulation in the immediate vicinity of the roots. This is important when irrigating salinized soils or irrigating with saline water.
- Drip irrigation allows the use of minimally treated sewage water because the water is delivered directly to the ground, minimizing health risks.
- High-quality drip irrigation equipment can last for fifteen to twenty years if handled properly.

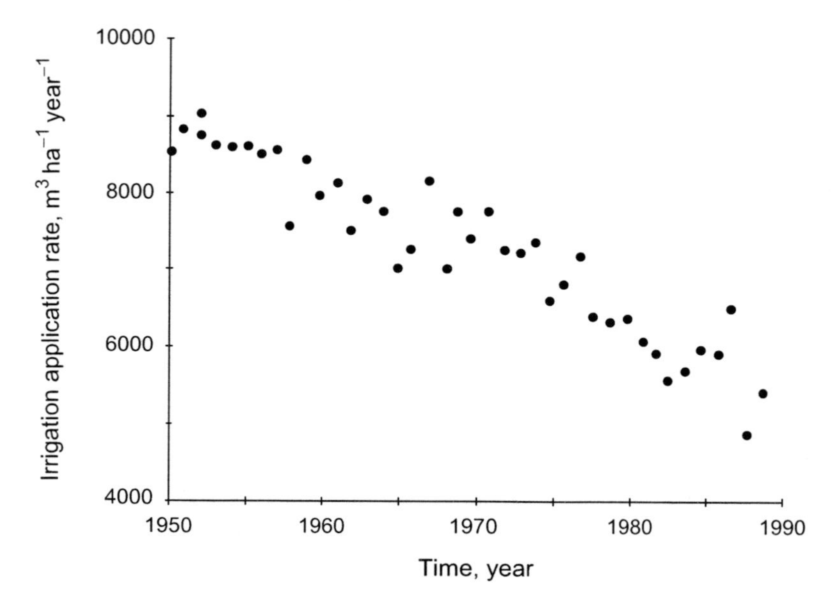

Figure 4. Average agricultural water use per hectare in Israel (Stanhill,1992)

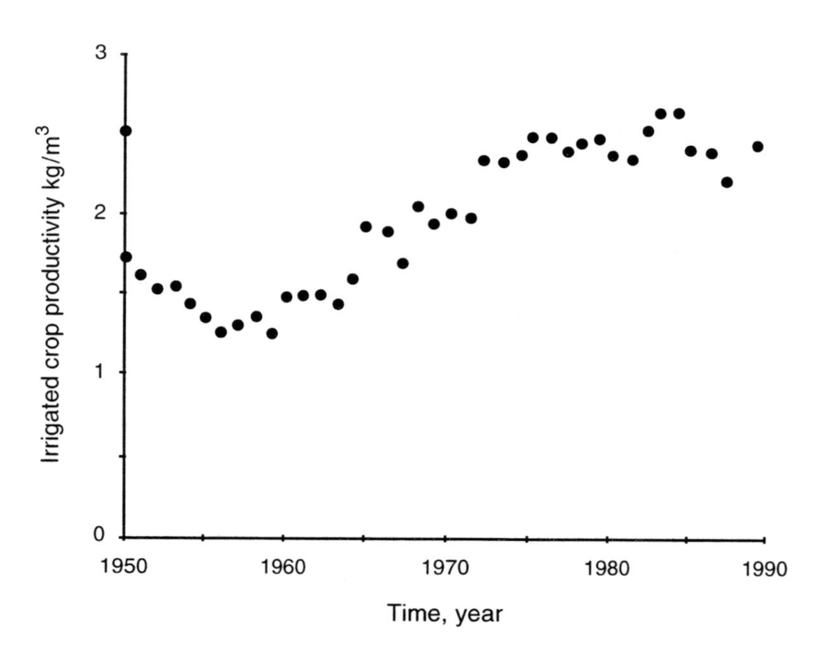

Figure 5. Irrigation crop productivity in Israel (Stanhill,1992)

- Using computer-controlled drip "fertigation" (applying fertilizer with the irrigation water) economizes on water and fertilizer use, and prevents soil salinization and groundwater pollution.

In sum, water use efficiency is about 45% in surface irrigation and 75% in sprinkler, while in drip irrigation it is about 95%. In addition, automated irrigation has resulted in better water control and the ability to irrigate at will, thereby reducing water losses. Consequently, it may be concluded that drip irrigation has many advantages over other methods of irrigation and that it is significantly superior to surface and sprinkler irrigation in regard water saving, especially under conditions of limited water supply.

The wide practice of a drip irrigation in Israel has permitted a lowering of the average agricultural water use per hectare from 8,600 m^3/yr in 1955 up to 5,700 m^3/yr in 1995 (Fig. 4) while crop productivity per unit of water (Fig. 5) increased more than twofold, from 1.2 to 2.5 kilogram per cubic meter (Stanhill, 1992). Israeli agricultural technology is illustrated in Plates 1–4.

5. Controlled environment agriculture

Greenhouses are very attractive option for hot deserts as, in comparison to conventional agriculture, they require very little fresh water or fertile soil, and can provide sophisticated, profitable employment for a modern

Plate 1: Greenhouse agriculture is an important branch of the Israeli economy.

a

b

Plate 2: a, b The Yair Research and Development Station in Arava valley is one of the most advanced stations of its kind, developing new strains of vegetables, new crops, organic crops, aquaculture, biological pest control and more. The Station also counsels local farmers with analyses of economic feasibility.

a

b

Plate 3: a, b The Yair Research and Development Station

a

b

Plate 4: a) Growing olive trees under drip irrigation by brackish water. Polyethylene film is used for reducing evaporation. Ramat Negev Desert Agricultural Experimental Center; b) Vegetables growing under drip irrigation.

farmer. Greenhouses in hot deserts face two main problems, cooling during periods of high insolation and heating during cold nights. The problem of overheating is solved either by ventilation and perhaps pad and fan cooling, or by limiting the use of the greenhouse to no more than six or seven months a year. In greenhouses energy costs are very low and there is a favorable combination of year-long controlled temperatures, filtered radiation, and high humidity. These conditions have resulted in growth rates two or three times higher than those obtained in the best conventional greenhouses, while maintaining or often improving the market quality (Gale et al., 1991). The total area of greenhouses an Israel reached 2,200 hectares in 1994. Many farmers attain record yields: for example, tomato plants are produced individually and can develop as creepers reaching 15 meters in length and producing up to 320 tons/hectare. Israel depends on irrigation and fertilization to increase crop yields. The country meets most of its food requirements through domestic production, using over a million cubic meters of water and about 95,000 tons of fertilizers annually. As a result, about 5 million tons of field crops, 1.2 billion liters of milk, 1.6 billion eggs, and 1.2 billion flowers are produced annually (The Environment in Israel, 2002).

6. Organic agriculture

The development of organic agriculture promises further reductions in environmentally harmful agricultural practices. Organic agriculture is a method of growing plants and raising livestock at their natural pace, with full consideration for life processes in their growth environment, and careful attention is paid to recycling and prevention of pollution in the environment. Similarly, pesticides, chemical fertilizers and hormones are not used.

Organic agriculture has increased rapidly in recent years, due to the demands of consumers for clean and ecological products. The product basket includes all types of vegetables, fresh fruit, dairy products, eggs, meat and poultry, as well as various processed products.

In 2003 organic agriculture included some 450 farmers who grew fresh produce on some 6,000 hectares (Statistical Abstract of Israel, 2004). Most of these products were intended for export. Theses farms are concentrated mainly in the northern and in the southern regions of Israel, because the central region is more affected by crop-dusting and other sources of pollution.

7. Afforestation

When Israel was established in 1948, there were fewer than 5 million trees in the entire area. Today, over 200 million trees have been planted in an active reforestation program. The program designates 160,000 hectares for the development and conservation of forest land in Israel – 7% of the total area of the country and over 15% of the land area of the north part of the Negev.

In Israel in 2002 there were 98,300 hectares of forest area or about 4% of the total land area (Statistical Abstract of Israel, 2002).

In the northern fringes of the Negev on the edge of the desert, Israel's largest man-planted forests serve as popular nature recreation venues. These forests rely on advanced water harvesting techniques which capture runoff rainwater in ridges, depression, terraces and limans. Single tree planting is yet another technique used to combat desertification in areas with even harsher natural conditions.

In light of the fact that less than 50% of rainwater in the Negev permeates into the underground water table, and most flows down in the gullies into the sea, new methods have been developed for flood prevention and water conservation. These involve the construction of small dams and trenches to collect and make optimal use of rainfall and storm runoff. They are used for growing trees and shrubs, for halting desertification of the Negev, and for directing rainwater into channels that recharge aquifers and create tourism and recreation sites. Runoff and rainfall catchment basins have enabled the development of runoff agroforestry and farming in areas with less than sufficient rainfall. Currently there are 12,000 hectares of established forest in the desert, with 200–300 hectares added each year (The Environment in Israel, 1994).

Man-made forests on previously barren slopes prevent soil erosion, promote biodiversity and restore limited productivity. Grasslands sprout around the widely spaced trees, providing pleasant scenery and limited grazing land. Run-off agroforestry provides food, firewood and fodder with a minimal investment in infrastructure.

8. Environmental protection and nature conservation

In a country as small as Israel, with a high rate of industrialization and urbanization, nature reserves are important to secure the biodiversity of the natural environment. Recognition of the need to protect Israel's precious natural and landscape resources led to the enactment of the national parks and nature reserves law in 1963. Israeli law defines a nature reserve as an area containing unique and characteristic animal, plant and mineral forms, which must be protected from any undesirable changes in their appearance, biological composition or evolution. Israeli reserves vary in size, character and use. Some encompass as little as one hectare while others span thousands of kilometers; most are open to the public and some offer special visitor services. Together, they represent the entire spectrum of Israel's ecosystems, including Mediterranean forests, marine landscapes, sand dunes, freshwater landscapes, desert and crater landscapes, and oases.

Plates 5–10 illustrate a selection of Israel's landscapes, plants and animals.

a

b

Plate 5: a) Liman is a man-made low-lying water reservoir dammed by dikes that trap runoff water. Trees planted in the liman depression survive on the gathered water and excess runoff flows through special channels to the wadi slope. Limans are built in the Negev areas where annual precipitation ranges between 50 to 300 mm; b) A liman in the Central Negev during spring.

a

b

Plate 6: The Negev in bloom: a) Tulips in Borot Lots; b) endemic *Iris yeruham* in the Yeruham national reserve

a

b

Plate 7: The Negev in bloom: a) Beeri natonal park; b) sun-rouse (*Helianthemum*) in Borot Lots (Photo Mori Chen)

a

b

Plate 8: a) Revival of the ancient water harvesting methods. Borot Lots area, Negev Highlands; b) The forest planted at the semi-arid part of the Negev, the Lahavr region

a

b

Plate 9: a) Antelopes in the Yotvata natural reserve "Hi Bar", Arava valley; b) A wady in the Arava Valley

Plate 10: Old *Pistacia atlantica* in a wady of the Negev Highlands

While nature reserves are predominantly concerned with the conservation of nature, national parks are primarily dedicated to the conservation of heritage and archaeology. Under the law, a national park is defined as an area of natural, scenic, historic, archeological or architectural value which is protected and developed for recreational purposes. National parks play an important role in protecting the country's natural beauty from rapidly encroaching urbanization and restoring and maintaining antiquities that have been lost or neglected for centuries (The Environment in Israel, 2002).

To date, 142 nature reserves and 44 national parks have been declared and established throughout the country (out of a total of 374 nature reserves and 113 national parks that are in various stages of planning). The statutory breakdown of the country's nature reserves and national parks in presented in the Table 1.

While awareness of the need to protect natural and landscape resources has led to the emergence of a significant system of nature reserves and national parks, the small size of the country and heavy pressure on its limited land resources have left few land reserves. As a result, protected areas are insufficient to preserve the natural values, the ecosystems and the unique landscape image of this highly diverse country. In order to secure biodiversity

TABLE 1. Statutory status of nature reserves and national parks (The Environment in Israel, 2002)

	Declared		Approved		Deposited		Proposed		Total	
	No.	Area km^2	No.	Area km^2	No.	Area km^2	No.	Area km^2	No.	Area km^2
Nature Reserves	142	3,224	29	752	6	11	197	1,734	374	5,720
National Parks	44	152	26	58	1	4	42	51	113	266

and visual resources, the country has formulated a new approach to development in open space landscapes which have not been designated as protected. This new approach seeks to direct development, both in terms of location and features, to appropriate areas in ways which will not destroy the ecosystem, the wildlife and the landscape features of each of the small but diverse landscape units in Israel.

Open space throughout the country was classified into four categories according to their value, importance, sensitivity and vulnerability: protected areas, open space landscape areas, controlled development areas, building and development areas.

Protected areas are areas of special importance and high sensitivity in relation to natural, landscape and historical values. Open space landscape areas are characterized by landscape sensitivity and are important for the protection of natural landscape diversity and features and for recreational needs. Controlled development areas are areas of intermediate natural and visual sensitivity, which are partly appropriate for building and development as long as landscape protection is taken into account. Finally, building and development areas are of low landscape sensitivity and are appropriate for building and development.

Environmental education and information are essential components in environmental programs. Their major objectives are to increase the environmental awareness of both the general public and decision makers, to educate them toward responsibility and concern for the environment, and to arouse their willingness and ability to contribute to environmental enhancement. It is widely recognized today that environmental awareness and understanding are prerequisites for environmental improvement.

In recent years, heightened concern about environmental issues has resulted in increased activism among the populace. New non-governmental environmental organizations are being created on the national level, while grass-roots groups are organizing in many areas to pressure authorities to seek solutions to environmental problems at the local level.

9. Conclusion

Israeli programs for combating desertification are based on sustainable agricultural development through centralized national water management that includes transportation of water from regions of water abundance to regions of water shortage, storage during years of abundance for use in years of drought, and cultivation of crops adapted to different water qualities and to the specific local climate and soil conditions,

Israel's agriculture is characterized by a high technological level and is based on drip irrigation systems, automatic and controlled mechanization, and high quality seeds and plants. The government is dedicated to increasing the efficient use of water and effluents and promoting water recycling, minimizing pesticide use, advancing organic agriculture, and development new crops and innovative machinery.

Israel pioneered forestry and forest rehabilitation methods for drylands. Using run-off harvesting technology, Israel succeeded in afforestation of regions with 200mm of rain per year. Israel established a preliminary template for a national action plan on desertification, with emphasis on research. This plan compiled a desertification research program that includes 50 multidisciplinary project proposals. The list of urgent activities that may constitute a framework for a national action plan includes actions for assessing, combating and monitoring soil salinization and sheet and gully erosion and for improving the management of rangeland, woodland fires, and road construction and use. Major emphasis will be placed on increasing the awareness of the public and decision-makers of the risk and damages of desertification.

The high priority accorded by Israel to the global effort in combating desertification led to the establishment of an International Center for Combating Desertification at the Ben-Gurion University campus in Sde-Boker, located within the arid region with 100mm of rainfall. The Albert Katz International School of Desert Studies is the major instrument of the International Center for disseminating know-how and transferring technologies for sustainable dryland development to trainees, students and experts from different countries.

References

Fisher, J. T., et al., 1999, Land use and managements: Research implications from three arid and semiarid regions of the world, Arid Lands Managements, pp. 149–170.

Gale, J., Richmond, A., and Appelbaum, S., 1991, Desert biosystems, in: *Combating Desertification and Improving the Quality of Life: The Israeli Case*, Sde-Boker, pp. 43–48.

Kliot, N., 1994, *Water Resources and Conflict in the Middle East*, Routledge, London, pp. 237.

Lithwick, H., 1998, Evaluating water balances in Israel, Negev Center for Regional Development, Beer Sheva, **10**: 37.

Nativ, R., and Iassar, A., 1989, Problems of an over-developed water system – the Israeli Case, in: Workshop Luso – Israelita sobre a agricultura intensive em zonas aridas e smi-aridas, Lisboa, pp. 169–184.

Shanan, L., 1998, Irrigation development: proactive planning and interactive management, in: H. Bruins, and H. Lithwick, ed., *The Arid Frontier*, Dordrecht, Kluwer Academic Publishers, pp. 251–276.

Sitton, D., 1998, *Development of Water Resources*, Israel Information Center, Ahva, Jerusalem, pp. 10.

Stanhill, G., 1992, Irrigation in Israel: past achievements, present challenges and future possibilities, in: *Water Use Efficiency in Agriculture,* J. Shalhevet, et all, eds., Rehovot, Priel Publisher, pp. 63–77.

Statistical Abstract of Israel, 2000, 2002, 2004, Central Bureau of Statistics, Jerusalem.

The Environment in Israel, 1992, 1994, 1999, 2002, Ministry of Environment, Jerusalem.

CHAPTER 10

POTENTIAL EFFECTS OF GLOBAL WARMING ON ATMOSPHERIC LEAD CONTAMINATION IN THE MOUNTAINS

GLOBAL WARMING AND LEAD CONTAMINATION

MARIÁN JANIGA*

Institute of High Mountain Biology, Zilina University, SK 05956 Tatranská Javorina 7, Slovak Republic

Abstract: Atmospheric metal contamination, especially lead, responds to both the environmental factors and global warming parameters as a function of altitude. Seasonal fluctuations in the lead concentrations of foliar parts of alpine plants have been recorded with values higher in winter and early spring months than in summer months. The larger the snow-free catchment area and the warmer conditions will be in the mountains, the larger and earlier dispersal of lead to the surrounding sub-mountain regions may occur. Pb and Al concentrations in the alpine plants and vertebrates must be of concern in acidified habitats.

Keywords: Atmospheric lead and aluminum, high mountains, *Rupicapra rupicapra*, climate change, acidification

1. Introduction

Currently, a large percentage of the scientific community have reached a consensus on various issues related to global warming. They are mainly: Changes in the fundamental physics of the Earth, added greenhouse gases add heat, the gases are anthropogenic, their reduction will require many years and the full recovery will require many centuries, reduction of sea ice is increasing, sea levels in general rise, global climate change will be greatest at higher latitudes, an increase in the amount of UV-B radiation reaching the earth's surface will occur, large stratospheric cooling will increase, and precipitation will increase, too (Martens and Rotmans, 1999; Jochem et al., 2001; McLaren and Kniveton, 2000; Visconti et al., 2001).

*To whom correspondence should be addressed. Institute of High Mountain Biology, Zilina University, SK 05956 Tatranská Javorina 7, Slovak Republic; e-mail: janiga@utc.sk

R. Behnke (ed.), The Socio-Economic Causes and Consequences of Desertification in Central Asia, 231–247.

Mountain regions cover about one-fifth of the continents and greatly influence regional and continental atmospheric circulation as well as water and energy cycles. Many studies predict that the effects of global climate change will be very intensive, both in terms of magnitude and speed of response, at higher altitudes (Price, 1999). Over the past three decades, significant efforts have been devoted to monitoring the effects in the climate of mountains; the most important changes seem to be:

- Extensive deglaciation in the world mountains in the last fifty years (e.g. Matthews, 1992)
- Precipitation amount normally increasing with elevation influences the effects of runoff, the effect being greater the higher the transient snowline, the larger the snow-free catchment area, and the higher the freezing level in atmosphere which may cause the catastrophic flooding (Collins, 1999)
- In Mediterranean and "dry" mountains, the contrast in streamflow between the wettest and driest years may be very large; evapotranspiration is usually the highest in the wettest and least in the driest years but when expressed as a percentage of rainfall the relative evapotranspiration is higher in dry than in the cold years (Coelho et al., 1999)
- The hydrology of mountains - "water towers of humankind" is gradually changing (Mountain Agenda, 1997)
- Plant successions at all vegetation levels of mountains are altered (Matthews, 1992); mainly alpine flora is currently moving upward (Pauli et al., 1996)
- The combined effects of rising atmospheric CO_2 and global warming are likely to cause more pronounced changes in alpine vegetation (Korner, 1992; Korner and Diemer, 1994); this may be of a crucial importance in the tropical and subtropical mountains where the mountains are able to absorb high amounts of CO_2 industrial emissions (Lasco and Pulhin, 1999)
- The functional linkage between the structure of alpine plant and invertebrate communities will bring new and very intensive changes in alpine ecosystems. Different species of alpine invertebrates respond to climate change in different ways and at different rates, mainly in relation to summer temperature and water availability (Hodkinson and Bird, 1998)
- Global warming may increase the sensitivity of alpine animals and people to increased acidity and heavy metal pollution in the high mountain environment

2. Metals in the mountains and global warming

Many studies have since confirmed that heavy metals travel in the atmosphere for thousands of kilometers within and across national borders, interacting with each other and forming secondary pollutants before being deposited.

Of these metals, depositions of Zn, Cd, Pb, Al and Hg are of an order of magnitude greater than those of the other metals (Scheuhammer, 1987, Steinnes et al., 1988). At high temperatures, metals are taken up in vertebrates much more rapidly by the intestine and are thus transported and accumulated rapidly in the tissues of the body (Wood et al., 2001).

2.1. MERCURY

Pollution must be tackled before global warming exacerbates its noxious effects. The metal works its way into the food chain, with women and children most at risk from poisoning, which can cause brain and nerve damage resulting in impaired coordination, blurred vision, tremors, irritability and memory loss. UNEP's (2005) report into the global impact of mercury pollution said more than 1,500 tones of the hazardous substance is pumped into the skies every year by power stations, with Asia and then Africa the worst culprits. Things could get worse in the coming years, as increases in temperature also appear to help the spread of the metal.

2.2. CADMIUM

A change in climate may lead to increased mobilization (weathering) of sites such as superfund sites and some agricultural soils that have become contaminated with cadmium. Increased exposure to UV may increase the genotoxicity of cadmium on skin cells. Tropical, arctic and alpine ecosystems are particularly sensitive to heavy metal pollution and because of the unique features of the food web, top predators of the ecosystems are more vulnerable to heavy metals compared to temperate species, and global warming can further exacerbate the exposure of the most sensitive organisms to cadmium and other toxic metal pollutants (Nriagu, 1996). Cadmium may increase damage of mitochondria in cadmium-exposed invertebrates with increasing temperature. These organelles become significantly more sensitive to cadmium as temperature rises, so that cadmium levels which were not damaging to mitochondria at lower temperature may become strongly toxic with increasing temperature (Sokolova, 2004).

2.3. ZINC

Has many of the same characteristics as lead with regard to deposition levels, deposition patterns (Berg et al., 1995) and emission levels (Pacyna, 1995). Zinc has been shown to antagonize a number of toxic effects of cadmium and lead by reducing their tissue accumulation and by reactivating lead-inhibited ALAD, an enzyme involved in heme synthesis (Hutton, 1983). Zn concentrations tend to increase with altitude (Šoltés,1998).

2.4. ALUMINIUM

Toxicity is primarily a function of the ability of Al to disrupt the dietary absorption and normal metabolism of Ca. Acidification causes increased environmental mobility of the metal (Scheuhammer, 1987, 1991a) Fish, alpine plants (Figure 1), invertebrates, and many vertebrate species (amphibians) are directly exposed to increased concentrations of Al, as a result of the acidification of their habitats (Scheuhammer, 1991b). Compared to the extensive literature on Hg or Cd, there is a dearth of information concerning the global warming (temperature increasing) and metal exposure effects on biota in the mountains. In alpine plants, the amount of Al tends to decrease in summer (but mean values were not statistically significantly different, Table 1) despite the high monthly amounts of rainfall in this period. Food

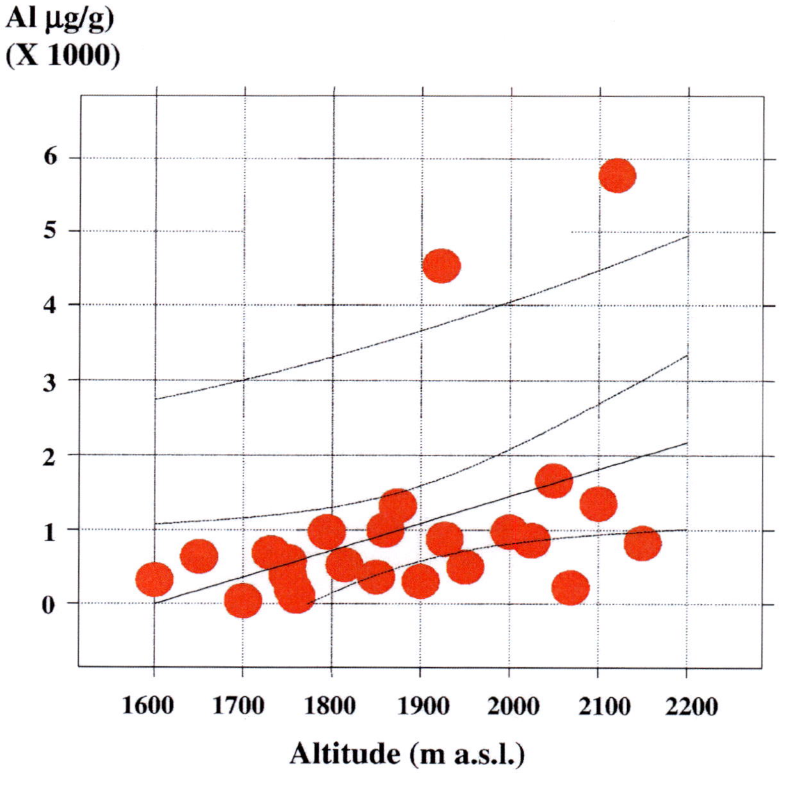

Figure 1. Aluminium concentrations in the food of chamois increase with altitude (linear regression model with 95 percent confidence and predictions limits: $y = -5803.2 + 3.63x$, $r = 0.42$, one-way ANOVA: $F = 5$, $P = 0.03$*, $n = 24$. Modified from Janiga (2004)

TABLE 1. Mean (SE) aluminum concentrations (μg/g dry weight) in chamois (*Rupicapra rupicapra*) food divided according to season, years of collection, location in the West Carpathians, and the type of the diet (vascular versus non-vascular plants; mainly lichen

Variable*	N	X	SE	MAX	MIN
Nov. - Febr.	11	1039.1	386.7	4530	81.5
March-April	9	1383.3	568.2	5770	302
June-August	6	617.5	268.8	1800	50.7
Sept.-Oct.	19	1038.1	459.7	9030	38.5
2002	26	804	184		
2003	19	1389	514		
West Tatras	10	1159	729		
Cetral Tatras (east)	12	1434	530		
Low Tatras	23	789	185		
Vascular plants	10	1941.8	955		
Non-vasc. plants	15	676	111		

*Within each of the groups, no significant difference exist by one-way ANOVA or Mann-Whitney rank sum test (types of the plants) at $p < 0.05$. Source: Janiga (2004)

of chamois *(Rupicapra rupicapra tatrica)* was not significantly affected by west-east location nor by the conditions in different years. The high mean values in the central Tatras reflect a trend of increasing Al pollution at higher altitudes. The vascular and non-vascular (eaten in winter) plants did not differ in aluminum levels. Therefore it is reasonable to assume that mountain ruminants are exposed to aluminum mainly in the early spring at highest sites of mountains.

3. Long-range transboundary lead pollution

Although the emissions of Pb to the global atmosphere have rapidly declined from 1983 to 1995, they are still on the order of from 27 to 45 times natural levels (Shotyk and Le Roux, 2005). These calculations apply to global scale emissions, so regional and local impacts must be even greater. Currently, the major source regions in Europe are to be seen in the densely populated and heavily industrialized regions of Central and East Europe. The most important anthropogenic source of Pb to the global atmosphere remains vehicle emissions; this source alone exceeds the natural fluxes by a factor of approximately 30 times. Other important sources of atmospheric Pb are non-ferrous metallurgy and fossil fuel combustion from stationary sources (mainly coal-burning) Prior to the introduction of leaded gasoline, 75% of the anthropogenic Pb inventory was probably already in the recent soils (mesured in the bogs (Shotyk et al., 2000). In contrast to declines in emissions from many sources of primary and secondary Pb, the emissions from stationary fossil fuel combustion tend to increase. Moreover, many nations in Asia, South America, Africa, and the Middle East have not yet banned the use of leaded gasoline. The largest primary producers of Pb are Australia (19%), the US (13%), China (12%), Peru (8%), and Canada (6%). The main use for Pb today (74%) is lead-acid batteries, especially for cars (Shotyk and Le Roux, 2005). Lead emitted from these sources is released to the air in the form of sub-micron particles, with a median diameter of 0.5 mm. The particles are considered to increase in size rapidly in atmosphere by coagulation with other particles. Due to their longer atmospheric residence time, smaller particles can be carried by winds and deposited over very large areas (Chamberlain et al., 1979).

4. Atmospheric depositions, orography and lead in soils

Trace elements are removed from the atmosphere by dry deposition (sedimentation, interception and impaction) and by wet deposition (rainout, washout). The first process is strongly dependent on particle size, wind velocity and surface characteristics, precipitation scavenging is largerly a measure of the liquid water content in the precipitation clouds (UNECE, 1995). Dry deposition may be very significant for large waterbodies (Pain, 1994). Because Pb has to be transferred from the solid phase to the aqueous phase before it can be taken up by plants and aquatic organisms, the concentration of Pb in natural waters can be a sensitive indicator of the potential for Pb to become biologically availabe. For example, it is evident that the geographic pattern of lead concentrations in the bones of Tatranian chamois reflects the climatic conditions in the Tatras (the Carpathians). The observed fall in lead levels in animals from the West Tatras to the eastern Belianske Tatras supports the hypothesis that atmospheric depositions of lead may influence the accumulation of lead in animals in remote areas (Janiga et al., 1998). Comparable studies were done in northern Europe, where, for example, Kalas and Lierhagen (1992) showed that herbivorous animals, hare *Lepus timidus*, black grouse *Tetrao tetrix*, and willow grouse *Lagopus lagopus* had similar gradients with a tenfold increase in liver lead concentrations from Northern to polluted Southern Norway. Froslie et al., (1984, 1985) also found higher lead levels in ruminants in Southern Norway than in reference areas, and their data correlated well with lead atmospheric deposition patterns.

Lead is relatively immobile in soils, and virtually all of the anthropogenic Pb could be found in the topmost 20 cm with the downward migration velocity of Pb from 1 to 8 mms/year. Acidic soils have the greatest potential for Pb migration (Shotyk and Le Roux, 2005), The Pb isotope data from acidic soils in southern Sweden suggest that Pb is migrating and, along with Al and Fe, accumulating in the B horizon of podzols (Renberg et al., 2002). The fate of lead in soil is a response to a complex set of parameters including soil texture, mineralogy, pH and redox potential, hydraulic conductivity, abundance of organic matter and oxyhydroxides of Al, Fe, and Mn, in addition to climate. There is often strong relationship between lead concentration in soil and parent material (Sposito, 1989).

5. Lead in the mountains

There is evidence of increased deposition of the metals over large mountain areas of Europe. The major source regions are found in central and eastern Europe (UNECE, 1995). Investigations have demonstrated that there is a high level of pollution by lead in alpine regions, i.e. the type of ecosystems which substantially form the national parks and reserves throughout the world. For lead deposition in the mountains, the importance of the proximity of the source is known from many studies. For example, the cumulative mass of atmospheric, anthropogenic Pb was also calculated in the Swiss Alps, using peat cores from eight mires (Shotyk et al., 2000). The lowest values were found in the north-east part of the Swiss Alps and the highest values originated from the south side with direct exposure to the highly industrial region of northern Italy. The higher levels of lead were also found, for example, in the bog at Bagno in Ukraine laying westerly of the East Carparthian Mountains, and directly exposed to the industrial regions of the former Eastern Europe, compared to the other bogs from the same country (Shotyk and Le Roux, 2005). Comparable findings were obtained from the rural Sierra Nevada Mountains of California (Patterson, 1981; Shotyk et al., 2002). In alpine habitats the relative importance of more poisonous organolead increases. From this point of view, consideration must be given to altitudes above the planetary boundary layer (ca. 1,500 m) which are more influenced by long-range transport processes (Shotyk et al., 2002).

Mosses, lichens are very suitable organisms for biological monitoring of air pollution due to their specific physiological features which easily enable the measurement of heavy metal deposition (Zechmeister, 1995; Šoltés, 1998). Bryophytes have no roots, but instead rely exclusively upon atmospheric inputs for nutrient elements. They also receive Pb from the air, and retain it efficiently, allowing moss analyses to be used as a monitoring tool for studying changes in atmospheric metal deposition. The isotopic composition of Pb in forest moss species such as *Hylocomium splendens* and *Polytrichum formosum* which have been collected annually during the past decades, allows a reconstruction of the predominant sources of anthropogenic Pb and their temporal variation. Lichens can be used for the same purposes, but because they grow on a rock substrate, more care might be needed to separate atmospheric from lithogenic inputs (Shotyk and Le Roux, 2005). Herbarium samples for *Sphagnum* moss which had been collected from peat bogs in Switzerland since 1867 (Weiss et al., 1999) and Scotland since 1830 (Farmer et al., 1999) may document temporal changes in the isotopic composition of atmospheric Pb over longer time periods. Ombrotrophic bogs are excellent archives of atmospheric Pb deposition because they receive Pb only from the air, and because they efficiently retain this metal despite the low pH of the waters, the abundance of natural, complexformingorganic acids and the seasonal variations in redox potential. Bogs are probably the best continental archives of atmospheric Pb deposition and they are receiving increasing attention for this purpose (Shotyk and Le Roux, 2005).

In vascular plants lead is absorbed mainly by roots and to only a limited extent by leaves, blades and peticles (Pain, 1994). Bednářová & Bednář (1978) experimented with many Tatranian vascular plants, and found that approximately 50% of lead is physiologically absorbed by the tissues of plants, while the others 50% may be washed away. Many alpine species of vascular plants as well as lichens and mosses contain elevated levels of lead in their tissues (Kyselová and Maňkovská, 1985; Šoltés et al., 1992), and some of the plant species with higher concentrations of lead in their tissues may be components in the diet of chamois (Table 2 and Figure 2).

Bengtsson and Tranvik (1989) reviewed literature on metal effects on soil invertebrates. Three main biological factors control metal accumulation in different groups of terestrial soil animals: the diet, the structure and physiology

TABLE 2. Mean (SE) lead concentrations (μg/g dry weight) in chamois (*Rupicapra rupicapra*) food divided according to season, years of collection, location in the West Carpathians, and the type of the diet (vascular versus non-vascular plants; mainly lichens)

Variable*	N	X	SE	MAX	MIN
Nov. - Febr.	11	13.5	2.99	34.7	1.5
March-April	9	19.4	3.14	33.2	9.9
June-August	6	15.0	5.88	35.0	1.9
Sept.-Oct.	19	11.0	1.33	23.2	2.1
2002	26	12.2	1.4		
2003	19	16.0	2.6		
West Tatras	10	17.5^{a}	3.1		
Cetral Tatras (east)	12	8.8^{b}	1.9		
Low Tatras	23	$14.9^{a,b}$	2.5		
Vascular plants	10	16.1	4.2		
Non-vasc. plants	15	16.4	1.8		

No significant differences were found among seasons (one-way ANOVA: $F = 1.81$, $P = 0.16$), between years ($F = 1.9$, $P = 0.17$), and types of diet ($F = 0.006$, $P = 0.94$); in localities, means with different letter indices in a column are significantly different by one-way ANOVA ($F = 2.9$, $P = 0.05$).
Source: Janiga (2004).

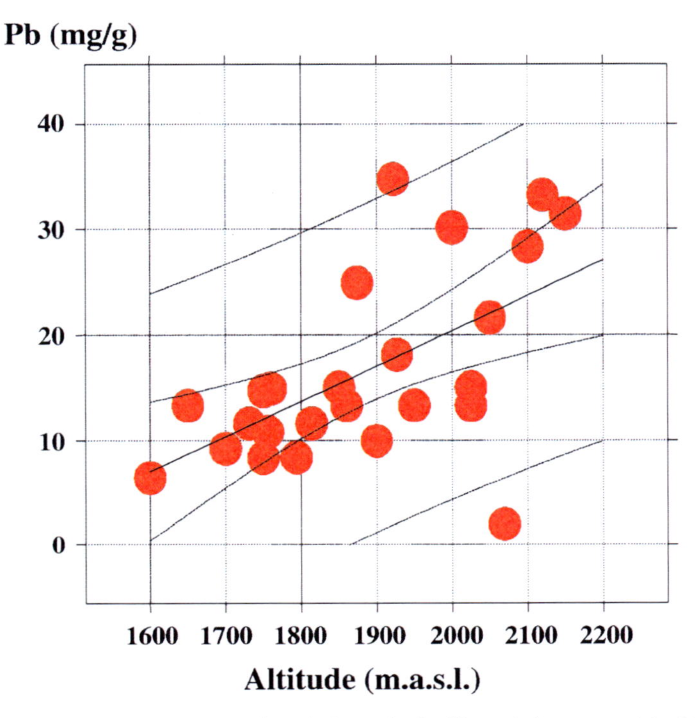

Figure 2. Lead concentrations in the diet of chamois significantly increase with altitude (linear regression model with 95 percent confidence and predictions limits: $y = -46.4 + 0.033x$, $r = 0.57$, one-way ANOVA: $F = 11.3$, $P = 0.003^{**}$, $n = 24$). Modified from Janiga (2004)

of digestive system, and the mechanisms by which the metals are stored. The soil communities may be strongly exposed to airborne heavy metals.

The effects of lead upon aquatic and semiaquatic organisms vary according to species and many hydrological, environmental and lead chemistry factors. In amphibians, lead exposure has resulted in a range of developmental and physiological effects (Pain, 1995).

Lead concentration levels in bird bones indicate that partially or fully granivorous birds have a higher degree of exposure to lead in the environment than other birds species (Scheuhammer, 1987). In mountains, of interest is the relatively high degree of lead exposure that crows, finches, thrushes, and especially alpine accentors have in the environment (Janiga, 2001). Lead concentration in the bones of the most applicable bioindicator - alpine accentor (*Prunella collaris*) reflected that lead was deposited in the alpine areas of the Tatra Mountains as long-range air pollutants. For granivorous and frugivorous birds which live in forest and rural areas, the high individual variation of lead in the bones probably reflected the local sources of lead polution. Lead was found at higher levels in granivorous and frugivorous birds than in birds predominantly eating invertebrates. Concentrations of lead in bones were significantly lower in birds of prey and owls.

Lead usually does not pose a serious hazard to adults of altricial passerines (Grue et al., 1984; Nybo et al., 1996; Arenal and Halbrook, 1997), but old-dying adults (Janiga and Žemberyová, 1998) and nestlings may exhibit a high sensitivity to lead. The poor breeding success of passerine birds in polluted areas is often related to the high amount of lead and other heavy metals in their diet (Bel'skii et al., 1995; Nyholm, 1995; Nyholm et al., 1995; Eeva and Lehikoinen, 1995). Heavy metal contamination of females of passerine birds affects the clutch size and embryo survival.

In mammals, children and other young animals are also more susceptible than adults to lead poisoning, due to increased lead uptake and incomplete development of metabolic pathways and the blood-brain barrier (Nyholm, 1994). In the Tatra chamois, there is a significant increase in bone lead concentrations with age. Embryos and kids contain less lead in their bones than yearlings and adults. The amount of lead in the bones dramatically increases in the first winter – spring period of life of juveniles (Janiga et al., 1998). Because the different seasonal types of the diet do not differ in the amount of accesible lead (Table 2), we assume that juveniles of chamois are also able to absorb a greater proportion of ingested Pb than do adults.

6. History of lead accumulation in the last century

In the West Tatras the level of lead did not change from 1946 to 1995 as might have been expected (Table 3). Lead levels found for the samples before World War II were also astonishingly high considering the small number of cars and the fact that addition of organolead compounds to gasoline started in the thirties only. The reason for the high levels is probably found in the

TABLE 3. Lead concentrations (μg/g dry weight) in the bone tissues of adult Tatranian chamois (*Rupicapra rupicapra*) from the different Tatra regions from 1905 to 1995

Variable	N	Median*	Interquart. range
Low Tatras, 1983–1987	4	7.7[a]	14.5
West Tatras, 1905–1946	17	11.7[a]	11.4
West Tatras, 1964–1974	7	16.8[a]	17.5
West Tatras, 1983–1987	7	15.8[a]	224.4
West Tatras, 1988–1995	4	15.7[a]	23.3
Central Tatras, 1988–1995	9	8.1[a]	5.9
Belianske Tatras,1964–1974	6	1.9[b]	1.0

*Different suffices in a column indicate statistically significant differences ($P < 0.05$, Man-Whitney rank sum test). Source: Janiga *et al.* (1998).

higher use of coal for general power production and heating purposes as coal contains varying portions of metals, including lead, which are released to the environment in smoke. The lack of filters resulted in high emissions of dust to which lead particles are adsorbed too and thereby dispersed around the source (Tataruch, 1995). Many lake sediments in Switzerland also contain elevated Pb concentrations at depths corresponding to the first few decades of the 20th century (Shotyk et al., 1998). The data from Denmark also suggest that coal burning was the main source of lead in this time. The lead examinations in the peat cores in Denmark reflect to a large extent the history of the coal industry and the chronology of the Second Industrial Revolution (Shotyk et al., 2003).

Since the Tatras represent a barrier for the pollutants from the northern part of central and east Europe (current major source of atmospheric lead in Europe), we suggest that this is the main reason why lead level did not decrease in the bones of chamois in the current times (Janiga et al., 1998). Vehicle traffic is a second important source of air pollutions. The rise of Pb in the bones of chamois from 1964 to 1974, is attributable to gasoline Pb that was used in many European countries. Since its introduction, gasoline Pb has dominated atmospheric Pb emissions in Euroasia, reaching a maximum between 1975 and 1982 (Shotyk et al., 1998). Samples of lead from chamois bones from 1983 to 1987 probably indicate that the West Carpathians were one of the most polluted regions by lead in Europe (Mirek, 1996). Although the lead compounds in the gasoline were substantially reduced from 1970 to 1990, by 1989 leaded petrol still contributed 69% of the total lead emissions in Europe (Nybo et al., 1996). While various sources of heavy-metal pollution (mining, refining, coal burning, gasoline Pb, and so on) are important Pb sources, soil dust deposition today remains an important component of the elevated Pb flux (Shotyk et al., 1998).

7. Potential effects of climate change

Demand for energy continues to rise, even in Europe. Increased use of fossil fuels threatens to accelerate climate change as well as metal pollution. In general, very incomplete information is available on field studies especially regarding chronic exposure of lead (months) to alpine biota. Little attention has been paid to synergistic or antagonistic effects of lead with the specific impacts of global warming.

In the Tatras (West Carpathians), it is generally known that there is a strong correlation between orography, the amount of wet deposition (Konček et al., 1973) and rainfall composition (Gazda and Hanzel, 1978). Precipitation rises more or less constantly with increasing altitude (Konček et al., 1973) and from easterly to westerly stations, being 1.7 times as high in the West Tatras as in the eastern Belianske Tatras (Petrovič and Šteffek, 1967). In the high altitude habitats, where chamois occur, the western and nortwestern winds mainly prevail. Moreover, windspeed may play an important role in the pollution of high altitude habitats (Clough, 1975). In the Tatras, wind speed increases above the tree line, mainly in the high altitude alpine areas. Deposition of lead probably initially depends on the element contents of air masses, and it is intensified by high windspeed in the alpine areas. The ratio of elements in the dust found in the Tatras corresponds to the ratio of elements found in some species of mosses (Šoltés et al., 1992), and in general the amount of lead in air dust tends to be larger than the amount of lead in the precipitation water (Tužinský and Chudíková, 1991; Šoltés et al., 1992). That means that particles of lead are probably brought by air masses – dry deposition. It is highly probable that dry deposition occurs from October to April when the rainfall is low in the alpine area (Smolen and Ostrožilík, 1994). Comparably in Arctic, there are pronounced seasonal variations in lead accumulation, and the snow samples containing the greatest Pb enrichments are from winter when the Arctic is dominated by air masses originating in Eurasia. Aerosols in the Arctic today are still highly contaminated by industrial Pb (Shotyk et al., 2003). Seasonal fluctuations in the lead concentrations of foliar parts have been recorded with values sometimes an order of magnitude higher in winter months than in summer months. Increased concentrations in winter are thought to be related to higher deposition rates resulting from increased thermal stability in winter, possible higher lead uptake by dead and decaying material due to cuticular breakdown increasing permeability and facilitating lead access, and slower winter – pasture turnover rates (Crump and Barlow, 1982; Ratcliffe and Beeby, 1980). Accumulation of lead in the layers of alpine snow and consequent snow melting are probably responsible for lead particles gathering on the surface of mountain plants. The highest contamination of chamois diet occurs in early spring (Table 2). In ruminants, there appears to be a seasonal occurrence of lead poisoning, usually in spring or summer (Osweiler et al., 1978). Then, from May to September, the lead may be washed from the rocks or plants to the alpine soil. Hajdúk (1988) found that lead content in the soil samples was higher close to rocky walls than in the samples at the distance from 2 to 8 meters from the wall. Lead content in the soil near the trunk of beech trees close to a gully was twice as high as concentrations on the opposite side of the trees. The author also found the increased elevations of soil lead in the northern valleys of the West Tatras (exposed to Polish coal fields). Rain and dew transport lead to soils by washing deposited lead off plant and rock surfaces; deposition tends to be lowest on smooth surfaces (Little and Wiffen, 1977).

In the second half of 20th century, the annual rainfall decreased from 2% to 12 % in the Tatra and Subtatra region but analyses of daily precipitation series show a trend in precipitation intensity and a tendency toward higher frequencies of heavy summer rainfalls in the last decades (Pacl, 1994). The larger the snow-free catchment area and the warmer conditions will be in Tatras, the larger and earlier the dispersal of lead to the environment. The increased amount of lead and decreased amount of water in the streams and rivers of mountain valleys may cause large-scale regional problems. Extensive work has been done to estimate how much of Pb might be transferred to the aqueous phase. Solutions collected from ten Swiss forest soils were measured for concentrations of total dissolved Pb and found to range from 1 to 60 mg/L. Five of the profiles were acidic (pH 4–5) and these contained 20–60 mg/L Pb in the aqueous phase of the surface layers. These layers, however, are the most critical ecologically, as they represent the biologically active zone of acidic forest soils; this is also the zone which has been most impacted by anthropogenic Pb (Shotyk and Le Roux, 2005). There is evidence to suggest that Pb or Al concentrations in the alpine plants and vertebrates must be of concern in acidified habitats and in instances where the availability of dietary Ca is low, i.e. in the granite parts of the mountains. If mountains consist of limestone, the effect is not so severe. The limestone helps to neutralize the acid. In granitic habitats, there is no such buffer. Acid solutions free nutrients as well as toxic metals from the soil. As a result, nutrients are lost from the soils and the plants may take on toxic elements. Exposure of wildlife to the potentially toxic lead must be of concern in instances where the availability of dietary Ca and P is low and the availability of the lead is high. Depressed levels of Ca are particularly effective in increasing the uptake and toxicity of Pb in birds and mammals (Scheuhammer, 1991a, 1991b; Eeva and Lehikoinen, 1995).

Effects of increased temperature and lead deposition influence many biological processes which may include:

- distribution and concentration of lead within the bodies of different organisms
- changed seasonality in the lead poisoning
- lead tolerance by plants and animals
- inhibition of heme-biosynthesis in vertebrates
- development of genetically tolerant organisms (strains) to lead contamined environment
- increased sensibility of nervous systems

- decreased availability of essential elements (e.g. Ca)
- change and decline in biodiversity in the sensitive regions (alpine habitats)

8. Conclusions

The results obtained in this work encourage additional research where the goals are to ascertain the effect of climate change on the environmental cycle of lead and other heavy metals. The generation and the testing of appropriate field methods must be developed in potential target regions. There is also a significant lack of comprehensive multi-disciplinary data for impact studies, which is one of the pre-requisites for case studies of impacts on natural or socio-economic systems. The changed cycle of accumulation of metals in agricultural soils of mountains gives rise to increased risk of toxic metals in human food. It should be taken into account that the biosphere is exposed to a multiple stressor syndrome and science is far from recognizing all possible interrelations between stressors.

References

Arenal, C. A., and Halbrook, R. S., 1997, PCB and heavy metal contamination and effects in European Starlings (*Sturnus vulgaris*) at a superfund site, *Bull. Environ. Contam. Toxicol*, **58**: 254–262.
Bednářová, J., and Bednář, V., 1978, Lead content in the plants of the Tatra National Park, *Zborník TANAP*, **20**: 163–175.
Bel'skii, E. A., Bezel', V. S., and Polents, E. A., 1995, Early stages of the nesting period of hollow-nesting birds under conditions of industrial pollution, *Russian Journal of Ecology*, **26**: 38–43.
Bengtsson, G., and Tranvik, L., 1989, Critical metal concentrations for forest soil invertebrates. A review of limitations, *Water, Air, Soil Pollut*, **47**: 381–417.
Berg, T., Royset, O., Steinnes, E., and Vadset, M., 1995, Atmospheric trace element deposition: principal component analyses of ICP-MS data from moss samples, *Environ. Pollut.*, **88**: 67–77.
Chamberlain, A.C., Heard, M. J., Little, P., and Witten, R. D., 1979, The dispersion of lead from motor exhausts, Proc. R. Soc. Discussion Meeting, Pathways of pollutants in the atmosphere, London 1977, *Phil. Trans. R. Soc. Lond. A.*, **290**: 577–589.
Clough, W. S., 1975, The deposition of particles on moss and grass surfaces, *Atmos. Environ.*, **9**: 1113–1119.
Crump, D. R., and Barlow, P. J., 1982, Fctors controlling the lead content of pasture grass, *Environ. Pollut.*, **B**: 181.
Coelho, C. O. A., Ferreira, A. I. D., Walsh, R. P. D., and Shakesby, R. A., 1999, Hydrological responses to forest land-use in the central Portuguese coastal mountains, in: *Global change in the mountains*, M. Price, ed., New York, London: The Parthenon Publishing Group, pp. 121–122.
Collins, D. N., 1999, Rainfall-induced high-magnitude runoff events in highly-glacierised Alpine basins, in: *Global change in the mountains*, M. Price, ed., New York, London: The Parthenon Publishing Group, pp. 40–42.
Eeva, T., and Lehikoinen, E., 1995, Egg shell quality, clutch size and hatching success of the great tit (*Parus major*) and the pied flycatcher (*Ficedula hypoleuca*) in an air pollution gradient, *Oecologia*, **102**: 312–323.

Farmer, J. G., Eades, L. J., and Graham, M. C., 1999, The lead content and isotopic composition of Br coals and their implications for past and present releases of lead to the U.K. environment, *Environmental Geochemistry and Health*, **21**: 257–272.

Froslie, A., Norheim, G., Rambaek, J. P., and Steinnes, E., 1984, Levels of trace elements in liver from Norwegian moose, reindeer and red deer in relation to atmospheric deposition, *Acta Vet. Scand.*, **25**: 333–345.

Froslie, A., Norheim, G., Rambaek, J. P., and Steinnes, E., 1985, Heavy metals in lamb liver: Contribution from atmospheric fallout, *Bull. Environ. Contam. Toxicol.*, **35**: 175–182.

Gazda, S., and Hanzel, V., 1978, Problems of conserving subterranean waters of the Tatra National Park from the aspect of the contemporary hydrogeological and geochemical knowledge, *Zborník TANAP*, **20**: 183–206.

Grue, C. E., O'Shea T. J., and Hoffman, D. J., 1984, Lead concentrations and reproduction in highway-nesting barn swallows, *Condor*, **86**: 383–389.

Hajdúk, J., 1988, Contents of Pb, Cd, As, Fe, Cr, Zn, Cu, Ca, Mg and S in TANAP soils in relation to the effect of industrial immissions, *Zborník TANAP*, **28**: 251–261.

Hodkinson, I. D., and Bird, J., 1998, Host-specific insect herbivores as sensors of climate change in arctic and alpine environments, *Arctic and Alpine Research*, **30**: 78–83.

Hutton, M., 1983, The effects of environmental lead exposure and *in vitro* zinc on tissue delta-aminolevulinic acid dehydratase in urban pigeons, *Comp. Biochem. Physiol.*, **74**C: 441–446.

Janiga, M., 2001, Birds as bio-indicators of long transported lead in the alpine environment in: *Global change and Protected Areas,* G. Visconti, M. Beniston, E. D. Iannorelli, and D. Barba, ed., London, Kluwer Publishers, pp. 253–259.

Janiga, M., 2004, Contamination of the natural trophic base of the Tatra chamois by lead and aluminium, Final Report, University Žilina, Tatranská Javorina.

Janiga, M., Chovancová, B., Žemberyová, M., and Farkašovská, I., 1998, Bone lead concentrations in chamois *Rupicapra rupicapra tatrica* and sources of variation, Proc. 2nd World Conf. Mt. Ungulates 145–150.

Janiga, M., and Žemberyová, M., 1998, Lead concentration in the bones o the feral pigeons (*Columba livia*): Sources of variation relating to body condition and death, *Arch. Environ. Contam. Toxicol.*, **35**: 70–74.

E. Jochem, J. Sathaye, and D. Bouille, ed., 2001, Society, behaviour, and climate change mitigation, Kluwer, Dordrecht, Boston, London.

Kalas, J. A., and Lierhagen, S., 1992, Terrestrial monitoring of ecosystems. Metal concentrations in the livers of hares, black grouse, and willow ptarmigan in Norway, Norwegian Institution for Nature Research, Trondheim, **137**.

Konček, M., Hamaj, F., Smolen, F., Otruba, J., Murínová, G., and Peterka, V., 1973, Climatic conditions in the High Tatra Mountains, *Zborník TANAP*, **15**: 239–324.

Korner, Ch., 1992, Response of alpine vegetation to global climate change, *Catena Suppl.*, **22**: 85–96.

Korner, Ch., and Diemer, M., 1994, Evidence that plants from high altitudes retain their greater photosynthetic efficiency under elevated CO_2 *Functional Ecology*, **8**: 58–68.

Kyselová, Z., and Maňkovská, B., 1985, Content of heavy metals in certain terricolous lichens in the Tatra National Park, *Zborník TANAP*, **26**: 153–160.

Lasco, R. D., and Pulhin, F. B., 1999, The mountain ecosystems of the Philippines: opportunities for mitigating climate change, in: *Global change in the mountains,* M. Price., ed., New York, London: The Parthenon Publishing Group, pp. 174–175.

Little, P., and Wiffen, R. D., 1977, Emission and deposition of petrol engine exhaust Pb. I. Deposition of exhaust Pb to plant and soil surfaces, *Atmos. Environ.*, **11**: 437.

P. Martens, and J. Rotmans ed., 1999, *Climate Change: An Integrated Perspective*, Kluwer, Dordrecht, Boston, London.

Matthews, J. A., 1992, *The Ecology of Recently-Deglaciated Terrain*, Cambridge University Press, Cambridge.

S. McLaren, and D. Kniveton, ed., 2000, *Linking Climate Change to Land Surface Change*, Kluwer, Dordrecht, Boston, London.

Mirek, Z., 1996, Anthropogenic threats and changes of the nature, in: *Nature of the Tatra National Park*, Z. Mirek, ed., Krakow, Zakopane: TPN, PAN, pp. 595–617.

Mountain Agenda, 1997, Mountains of the world. Challenges for the 21st century, Mountain agenda, Bern.

Nriagu, J. O., 1996, A history of global metal pollution, *Science*, **272**: 223–224.

Nybo, S. P., Fjeld, E., Jerstad, K., and Nissen, A., 1996, Long-range air pollution and its impact on heavy metal accumulation in dippers *Cinclus cinclus* in Norway, *Environ. Pollution*, **94**: 31–38.

Nyholm, N. E. I., 1994, Heavy metal tissue levels, impact on breeding and nestling development in natural populations of Pied Flycatcher (Aves) in the pollution gradient from a smelter, in: *Ecotoxicology of soil organisms*, M. H. Donker, H. Eijsackers, and F. Heimbach, eds, London: Lewis Publishers, pp. 373–382.

Nyholm, N. E. I., 1995, Monitoring of terrestrial environmental metal pollution by means of free-living insectivorous birds, *Annali di chimica*, **85**: 343–351.

Nyholm, N. E. I., Sawicka-Kapusta, K., Swiergosz, R., and Laczewska, B., 1995, Effects of environmental pollution on breeding populations of birds in southern Poland, *Water, Air and Soil Pollution*, **85**: 829–834.

Osweiler, G. D., Van Gelder, G. A., and Buck, W. B., 1978, Epidemiology of lead poisoning in animals, in: *Toxicity of heavy metals in the environment*, F. W. Oehme, ed., New York: Marcel Dekker, pp.143–171.

Pacl, J., 1994, Waters, in: *Tatra national park*, I. Vološčuk, T. Lominca, ed., Gradus, Tatra National Park, pp. 66–78.

Pacyna, J. M., 1995, Emissions inventory for atmospheric lead, cadmium and copper in Europe in 1989, Institute for Applied System Analysis, Laxenburg, IIASA WP-95–35.

Pain, D., 1994, Lead in the environment, in: *Handbook of ecotoxicology*, D. Hoffman, B. A. Rattner, G. A. Burton, and J. Cairns, ed., Boca Raton, Ann Arbor, London, Tokyo: Lewis, pp. 356–391.

Patterson, C. C., 1981, An alternative perspective—Lead pollution in the human environment: Origin, extent, and significance, Commission on Lead in the Human Environment, U.S. Natl. Acad. of Sci., Washington, D. C.

Pauli, H., Gottfried, M., and Grabberr, G., 1996, Effects of climate change on mountain ecosystems – upward shifting of alpine plants.

Petrovič, O., and Šteffek, M., 1967, Evaluation of measurements recorded on precipitations totalisators in the Tatra region, *Zborník TANAP*, **10**: 105–110.

M. Price, ed., 1999, *Global change in the mountains. Parthenon*, NewYork, London.

Renberg, I., Brännvall, M. L., Bindler, R., and Emteryd, O., 2002, Stable isotopes and lake sediments- a useful combination for the study of atmospheric lead pollution history, *Science of the Total Environment*, **292**: 45–54.

Ratcliffe, D., and Beeby, A., 1980, Differential accumulation of lead in living and decaying grass on roadside verges, *Environ. Pollut.*, **A**: 279.

Scheuhammer, A. M., 1987, The chronic toxicity of aluminium, cadmium, mercury, and lead in birds: A review, *Environ. Pollut.*, **46**: 263–295.

Scheuhammer, A. M., 1991a, Effects of acidification on the availability of toxic metals and calcium to wild birds and mammals, *Environ. Pollut.*, **71**: 329–375.

Scheuhammer, A. M., 1991b, Acidification – related changes in the biogeo-chemistry and ecotoxicology of mercury, cadmium, lead and aluminium: Overview. *Environ. Pollut.*, **71**: 87–90.

Shotyk, W., Weiss, D., Appleby, P. G., Cheburkin, A. K., Frei, R., Gloor, M., Kramers, J. D., Reese, S., and van der Knaap, W. O., 1998, History of atmospheric lead deposition since 12,370 14C yr BP recorded in a peat bog profile, Jura Mountains, Switzerland, *Science*, **281**: 1635–1640.

Shotyk, W., Blaser, P., Grunig, A., and Cheburkin, A. K., 2000, A new approach for quantifying cumulative, anthropogenic, atmospheric lead deposition using peat cores from bogs: Pb in eight Swiss peat bog profiles, *Science of the Total Environment* **249**: 257–280

Shotyk, W., and Le Roux, G., 2005, Biogeochemistry and cycling of lead, in: Biogeochemical Cycles of the Elements, A. Sigel, H. Sigel, and R. K. O. Sigel, ed., *Metal Ions Biol. Syst.*, **43**: 240–275.

Shotyk, W., Weiss, D., Heisterkamp, M., Cheburkin, A. K., and Adams, F. C., 2002, A new peat bog record of atmospheric lead pollution in Switzerland: Pb concentrations, enrichment factors, isotopic composition and organolead species, *Environmental Science and Technology* **36**: 3893–3900.

Shotyk, W., Goodsite, M. E., Roos-Barraclough, Heinemeier, J., Frei, R., Asmund, G., Lohse, C., and Stroyer, T. H., 2003, Anthropogenic contributions to atmospheric Hg, Pb and As deposition recorded by peat cores from Greenland and Denmark dated using the 14C ams "bomb pulse curve", *Geochim. Cosmochim. Acta*, **67**: 3991–4011.

Smolen, F., and Ostrožilík, M., 1994, Climate, in: *Tatra national park*, I. Vološčuk, ed., T. Lomnica: Gradus, Tatra National Park, pp. 50–65.

Sokolova, I. M., 2004, Cadmium effects on mitochondrial function are enhanced by elevated temperatures in a marine poikilotherm, *Crassostrea virginica* Gmelin (Bivalvia: Ostreidae), *Journal of Experimental Biology*, **207**: 2639–2648.

Šoltés, R., 1998, Correlation between altitude and heavy metal deposition in the Tatra Mts (Slovakia), *Biologia, Bratislava*, **53**: 85–90.

Šoltés, R., Šoltésová, A., and Kyselová, Z., 1992, Effect of emissions on non-forest vegetation in the High Tatra and Belianske Tatry, *Zborník TANAP*, **32**: 307–333

Sposito, G., 1989, *The Chemistry of Soils*, Oxford: Oxford University Press.

Steinnes, E., Frantzen, F., Johansen, O., and Rambek, J. P., 1988, Atmosferisk nedfall av tungmetaller i Norge, Norwegian State Pollution Control Authority, Oslo, **335/88**.

Tataruch, F., 1995, Red Deer antlers as biomonitors for lead contamination. *Bull. Environ. Contam. Toxicol.*, **55**: 332–337.

Tužinský, L., and Chudíková, O., 1991, Precipitation, gutter and gravitation water chemism in the TANAP forest ecosystem, *Zborník TANAP*, **31**: 97–107.

UNECE, United Nations Economic Commission for Europe, 1995, Task force on heavy metals emissions, State-of-the-Art Report, 2nd Edition, Prague.

Weiss, D., Shotyk, W., Gloor, M., and Kramers, J. D., 1999, Herbarium specimens of *Sphagnum* moss as archives of recent and past atmospheric Pb deposition in Switzerland: isotopic composition and source assessment, *Atmospheric Environment* **33**: 3751–3763.

G. Visconti, M. Beniston, E. D. Iannorelli, and D. Barba, ed., 2001, Global change and protected areas, London, Kluwer Publishers.

Wood, C. M., Morgan, I., Reid, S. D., and McDonald, D. G., 2001, Global warming and the rainbow trout: interactive effects of environmental contaminants and dietary ration on the physiology of a stenothermal freshwater fish, in: *Climate Change: Effects on Plants, Animals, and Humans*, C. L. Bolls, ed., University of Milan, Milan.

Zechmeister, H. G., 1995, Correlation between altitude and heavy metal deposition in the Alps, *Environ. Pollut.*, **89**: 73–80.

CHAPTER 11

THE INFLUENCE OF ENVIRONMENTAL FACTORS ON HUMAN HEALTH IN UZBEKISTAN

A. M. SHAMSIYEV*[1] AND SH. A. KHUSINOVA[2]

[1,2]*Samarkand State Medical Institute, A. Temur str., 18, Samarkand, Uzbekistan*

Abstract: The incidence of communicative and tropical diseases is high in the Aral Sea region of Central Asia, effecting 60–300 per 10,000 people. High levels of chlorides in drinking water, the high incidence of heart disease, and increased levels of Ca and Mg leading to biliary and renal calculosis are all problems. Morbidity rates for biliary calculosis have increased ten fold, rates for chronic gastritis four fold, renal diseases eight fold and arthrosis and arthritis by five and seven times. The incidence of acute respiratory diseases in Karakalpakistan varies from 46% to 52%, with the incidence of tuberculosis having doubled in the last ten years so that it is now three times the rate elsewhere in Uzbekistan. Complications in pregnancy affect roughly two thirds of all women who are exposed to pesticides, with increases in pregnancy induced hypertension, gestational anemia, spontaneous abortions and preterm deliveries.

Keywords: Human health, pollution, Karakalpakistan, Aral Sea crisis, Uzbekistan

Medical ecology examines the consequences of environmental conditions for human health. In Uzbekistan the range of potential ecologically related diseases is wide, including fluorosis and osteopathic diseases due to high levels of fluoride in drinking water and the environment and endemic goiter caused by lack of alimentary iodine. In recent decades the environmental situation has deteriorated markedly in the Aral region of Central Asia due to the drying

* To whom correspondence should be addressed. Samarkand State Medical Institute, A. Temur str., 18, Samarkand, Uzbekistan; e-mail:

R. Behnke (ed.), The Socio-Economic Causes and Consequences of Desertification in Central Asia, 249–252.

of Aral Sea. Today the dried bed of Aral Sea covers 28,000 squire kilometers, and every year about 75 million tons of sand and dust and 65 million tons of salts are spread into the atmosphere by the wind and dispersed as far as a thousand kilometers. The deteriorating situation in Karakalpakistan adjacent to the Aral Sea has led to a significant worsening of public heath due to poor quality drinking water (increased mineralization, chemical pollutants, etc.), increased dust in the atmosphere, and the pollution of soil, plants and nutrients with salt and chemicals.

The hot, continental climate of the southern Aral region contributes to the development of water-related diseases since water needs and water consumption increase in a hot climate. Nonetheless, only 68% of Karakalpakistan's population is supplied with piped water, the rest obtaining poor quality drinking water from open sources, which contributes to the spread of contagious diseases. The incidence of communicative and tropical diseases is high in the Aral Sea region of Central Asia, effecting 60–300 per 10,000 people. High levels of chlorides in drinking water, the high incidence of heart disease, and increased levels of Ca and Mg leading to biliary and renal calculosis are all problems. Morbidity rates for biliary calculosis have increased ten fold, rates for chronic gastritis four fold, renal diseases eight fold and arthrosis and arthritis by five and seven times. It is estimated that exogenous environmental factors now account for 68% of child morbidity, of which 33% is derived from substandard water supplies, about 19% can be attributed to contaminated air, and 16% is caused by pesticides.

Following the drying of the Aral Sea, salt dusts from the dried sea bed have been deposited over a territory of 150 to 200 thousand squire kilometers. Long-term inhalation of dusty air leads to the development of bronchial asthma, bronchitis and allergic diseases. The incidence of acute respiratory diseases in Karakalpakistan varies from 46% to 52%, with the incidence of tuberculosis having doubled in the last ten years so that it is now three times the rate elsewhere in Uzbekistan. Tuberculosis is especially prevalent in heavily degraded areas of Karakalpakistan and is strongly correlated with the level of pesticides in the surrounding environment.

Intensive use of pesticides on farms in Uzbekistan has continued for many decades. Between 1980 and 1995, 69 different kinds of pesticides totaling 32 million tons of active ingredients were used. Some of this material is still circulating in the environment and has entered the human food chain. Up to the present, foods of both vegetable and animal origin have been polluted with pesticides and their metabolites, some of them are more toxic for humans than the initial substances. In Uzbekistan 85% of the pesticides absorbed by people are ingested with food.

Thirty years of intensive cotton cultivation in Uzbekistan have also caused pesticide pollution of soils, the gradual drying of the Aral Sea, and declining supplies of safe water, especially in the rural areas. One of the pathogenic factors of agricultural manufacture is contact with pesticides, which leads to their intake and accumulation in the organism. There is considerable evidence of acute pesticide poisoning, but the influence of smaller amounts of pesticides is insufficiently studied and it would appear that small amounts of many different chemicals affect human reproduction. Oxygen is implicated in this process. In healthy persons the level of superoxide anions is low and their toxic effects are negligible. However, as liposoluble substances, pesticides generate increased levels of superoxide anions. These active forms of oxygen damage DNA molecules and are potentially mutagenic and carcinogenic. Peroxides are also damaging to the reproductive system. As a result, 63% of women in Uzbekistan exposed to pesticides were threatened by early termination of their pregnancies. Damage to the female reproductive system and elevated numbers of births with congenital abnormalities are closely related to pollution levels. The influence of the parents' occupation on the rate at which malignant tumors develop in children is evident even in cases when exposure occurred a year before than a baby was born. The frequency and severity of pregnancy induced pathologic conditions have also increased – that of hypertension by up to 7%, gestational anemia by 12%, and spontaneous abortions and preterm deliveries by 9%. The final results of those disorders are fetal and neonatal pathologies – preterm deliveries and underweight babies with asphyxia, hypotrophy and often congenital abnormalities. In Karakalpakistan more than 85% of women suffer from chronic kidney diseases. When the children of the region were examined almost all of them had anemia, 64% had second or third degree hypotrophy, 57% displayed physical growth retardation, and 10% had unfavorable obstetric anamnesis.

Diseases of digestive system are wide spread and often recurrent in individuals in the Aral region. Morbidity rates for gastrointestinal diseases are 74,9%, and gastritis, duodenitis and liver disorders are especially common. Ecologic factors also impair the capacity of patients to cope with diffuse liver diseases, exacerbate the development of acute viral hepatitis, increase the probability of it becoming chronic, especially in children and adolescents, and accelerate the development liver cirrhosis and its complications. There are strong statistical correlations between gastrointestinal disease morbidity and the amount of chlorides, sulfates, and the prevalence of pesticides in drinking water.

The water reservoirs serving Khoresm, Navoiy, Bukhara, Tashkent and the Ferghana contain high concentrations of fluoride, chrome, arsenic, ammonia and nitrates. Arsenic ingested in small amounts in water or food accumulates and causes chronic poisoning which increases the severity of chronic diseases and renders therapy more difficult. High concentrations of heavy metals in water and soil also alter soil microbial processes and present a cancer risk. The increased intake of zinc contributes to the growth of tumors and many kinds of cancer are related to cadmium intake. Increased level of cadmium causes bone abnormalities, and calcium insufficiency damages mineral exchange in bone tissues. Simultaneous iodine and manganese insufficiency appears with endemic goiter which is very common in the territory of Aral region.

Despite these health risks and severe diseases, many patients reject hospitalization, do not submit to medical examination and prefer self-treatment. But it must also be noted that it is necessary to better equip laboratories with modern instruments that will permit the more detailed examination of patients.

Thus, the pollution of the natural environment significantly influences health and disease rates among the population of Uzbekistan. The Aral crisis and its consequences on the surrounding environment and human health have few parallels world wide. This is a global as well as a regional Central Asian problem, and a challenge to international science which must be addressed before it is too late.